AF818815

Springer Theses

Recognizing Outstanding Ph.D. Research

Aims and Scope

The series "Springer Theses" brings together a selection of the very best Ph.D. theses from around the world and across the physical sciences. Each volume has been selected for its scientific excellence by an internationally renowned institute or department. For greater accessibility to non-specialists, the published versions include an extended introduction, as well as a foreword by the student's supervisor explaining the special relevance of the work for the field. The series as a whole provides a valuable resource both for newcomers to the research fields, and for other scientists seeking detailed background information on special questions. Finally, it serves as an accredited documentation of the valuable contributions made by today's younger generation of scientists.

How to Apply

Theses prepared at internationally leading universities and institutes may be nominated for publication by supervisors with the endorsement of the head of department. Each such department is entitled to nominate one thesis per year. Enquiries and requests for a nomination form should be sent, together with the thesis, to Ms. Nathalie Berg (nathalie.berg@springer.com).

Eligibility Criteria

To qualify for consideration, theses should fulfil the following criteria:

- They must be written in good English.
- The topic should fall within the confines of Chemistry, Physics, Earth Sciences, Engineering and related interdisciplinary fields such as Materials, Nanoscience, Chemical Engineering, Complex Systems and Biophysics.
- The work reported in the thesis must represent a significant scientific advance.
- A maximum of 30% of the text may be a verbatim reproduction from the author's previous publications.
- They must have been examined and passed during the 12 months prior to nomination.
- The theses should have a clearly defined structure including an introduction accessible to new PhD students and scientists not expert in the relevant field.
- After acceptance, theses should be supplemented by a "Supervisor's Foreword" outlining the wider scientific relevance of the work.

Indexed by zbMATH.

Tingzhao Fu

Research on Silicon-Based Integrated Diffractive Optical Neural Network

Doctoral Thesis accepted by
Tsinghua University, Beijing, China

清華大学出版社
TSINGHUA UNIVERSITY PRESS

Springer

Author
Dr. Tingzhao Fu
National University of Defense Technology
Changsha, Hunan, China

Supervisor
Prof. Hongwei Chen
Tsinghua University
Beijing, China

ISSN 2190-5053 ISSN 2190-5061 (electronic)
Springer Theses
ISBN 978-981-95-3786-0 ISBN 978-981-95-3787-7 (eBook)
https://doi.org/10.1007/978-981-95-3787-7

Jointly published with Tsinghua University Press
The print edition is not for sale in China (Mainland). Customers from China (Mainland) please order the print book from: Tsinghua University Press.

© Tsinghua University Press 2026

This work is subject to copyright. All rights are solely and exclusively licensed by the Publisher, whether the whole or part of the material is concerned, specifically the rights of translation, reprinting, reuse of illustrations, recitation, broadcasting, reproduction on microfilms or in any other physical way, and transmission or information storage and retrieval, electronic adaptation, computer software, or by similar or dissimilar methodology now known or hereafter developed.
The use of general descriptive names, registered names, trademarks, service marks, etc. in this publication does not imply, even in the absence of a specific statement, that such names are exempt from the relevant protective laws and regulations and therefore free for general use.
The publishers, the authors, and the editors are safe to assume that the advice and information in this book are believed to be true and accurate at the date of publication. Neither the publishers nor the authors or the editors give a warranty, express or implied, with respect to the material contained herein or for any errors or omissions that may have been made. The publishers remain neutral with regard to jurisdictional claims in published maps and institutional affiliations.

This Springer imprint is published by the registered company Springer Nature Singapore Pte Ltd.
The registered company address is: 152 Beach Road, #21-01/04 Gateway East, Singapore 189721, Singapore

If disposing of this product, please recycle the paper.

Supervisor's Foreword

Artificial intelligence is sweeping across the globe at an unprecedented pace, yet the underlying computational architecture that supports its development faces dual challenges of computing power and power consumption. Optical neural networks represent a highly forward-looking computational architecture, boasting significant advantages such as ultra-high speed, low-power consumption, and large bandwidth. Currently, there has been extensive research in areas such as image processing and speech recognition, positioning it as a strong contender for the next generation of computational architectures. Among these, diffractive optical neural networks (DONNs), characterized by their high integration and high computing capacity, hold the promise of serving as a bridge for the perfect combination of optical computing and artificial intelligence.

This book systematically investigates the fundamental principles, design method, simulation, and experimental validation of on-chip integrated DONNs, covering the complete chain from theoretical foundations to practical verification. The main innovations include: ① A parameter training and mapping model for on-chip DONNs is proposed, which can analytically characterize the propagation laws of light in silicon-based slab waveguides and enable more accurate implementation of pre-trained key parameters on physical devices; ② A silicon-based integrated DONN chip with low latency, low-power consumption, high integration, and high computing capacity is proposed, along with a compensation scheme for system errors in the chip, further ensuring the performance of the DONN chip system; ③ A design methodology for a multifunctional silicon-based integrated DONN chip system is proposed, which further enhances the computing efficiency of the DONN chip.

This research topic holds important theoretical significance and practical value. Most of the content covered in this book has previously been published in the foremost journals or conferences within this research field. Despite the rapid development in optical neural networks, understanding the fundamental technologies and principles of on-chip DONNs can provide readers with a comprehensive, systematic, and in-depth reference. For readers engaged in interdisciplinary research in optics, electronic information, artificial intelligence, and related fields, this book may help them apply similar techniques or inspire them to develop novel solutions.

Beijing, China
March 2025

Hongwei Chen

Related Publications in This Thesis

[1] Fu T, Zang Y, Huang Y, et al. Photonic machine learning with on-chip diffractive optics[J]. Nature Communications, 2023, 14(1): 70
[2] Fu T, Zang Y, Huang H, et al. On-chip photonic diffractive optical neural network based on a spatial domain electromagnetic propagation model[J]. Optics Express, 2021, 29(20): 31924–31940
[3] Fu T, Huang Y, Sun R, et al. Integrated diffractive optical neural network with space-time interleaving[J]. Chinese Optics Letters, 2023, 21(9): 091301

Contents

Acronyms

2.5D FDTD	2.5-Dimensional Variational FDTD
AM	Amplitude Modulator
ASIC	Application-Specific Integrated Circuit
AWG	Arbitrary Waveform Generator
CMOS	Complementary Metal Oxide Semiconductor
CPU	Central Processing Unit
DMD	Digital Micro-mirror Device
DONN	Diffractive Optical Neural Network
EMT	Effective Medium Theory
FA	Fiber Array
Fashion-MNIST	Fashion-Modified National Institute of Standards and Technology Database
FDTD	Finite Difference Time Domain
FPGA	Field-Programmable Gate Array
MMI	Multimode Interferometer
MNIST	Modified National Institute of Standards and Technology Database
ONN	Optical Neural Network
OPS	Operations per Second
PCB	Printed Circuit Board
PD	Photodetector
SDEPM	Spatial Domain Electromagnetic Propagation Model
SiO_2	Silicon Dioxide/Silica
SLM	Spatial Light Modulator
SOI	Silicon-on-Insulator
SSSD	Silicon Slot Filled with Silicon Dioxide
SSSDG	Silicon Slot Filled with Silicon Dioxide Group
TOPS	Trillions of Operations per Second
USPS	United States Postal Service

Chapter 1
Research Background and Progress

Abstract This chapter will briefly introduce the research background of Optical Neural Networks (ONNs), while also reviewing the developmental trajectory of ONNs. As a novel computing paradigm, ONNs possess inherent advantages such as low latency, low power consumption, and ultra-fast parallel signal processing compared to electronic neural networks. Although ONNs experienced slow development in the past due to limitations in fabrication technology, societal demands, and data resources, recent advancements in these areas have facilitated rapid progress in ONNs. This chapter will provide a detailed overview of the current research status of ONNs, analyze potential areas for improvement in different types of ONN studies, and present the research content, significance, and objectives to be explored in this thesis.

1.1 Introduction

Over the past two decades, the rapid development of semiconductor process technology [1] and the continuous optimization and improvement of computing hardware architectures, including GPUs, FPGAs, and ASICs [2–8], have made computing resources increasingly abundant. Driven by the vast computing resources, neural networks have achieved widespread application and remarkable progress across diverse domains [2–8], such as computer vision [9], natural language processing [10], speech recognition [11], autonomous driving [12], and biomedical sciences [13–15]. Nevertheless, as the appetite for intelligent solutions grows, the complexity of tasks that neural networks are expected to manage has escalated correspondingly, leading to a rapid surge in computing resource consumption, especially when tackling intricate tasks. By 2020, the precision of the semiconductor process had approached 3 nm [1], which is approaching the physical limit of transistors. Concurrently, the field of artificial intelligence is evolving at a breakneck pace, with its hunger for computing resources intensifying daily. In the future, as neural networks face even

© Tsinghua University Press 2026

T. Fu, *Research on Silicon-Based Integrated Diffractive Optical Neural Network*,
Springer Theses, https://doi.org/10.1007/978-981-95-3787-7_1

more complex challenges, the escalating demand for computing resources is anticipated to encounter a significant bottleneck. Therefore, a new computing paradigm is urgently needed to address and surmount this looming constraint.

In 1989, H. John Caulfield et al. systematically introduced optical neural networks (ONNs) [16]. They indicated that the mutual penetration of traditional optical information processing and neural network knowledge systems created ONNs and believed that, in certain specific cases, ONNs would have more significant advantages than electronic neural networks. At that juncture, ONNs were predominantly constructed with discrete optical elements, which limited their applicability. Moreover, the specific conditions, such as the absence of the current societal demand for substantial computing resources, the nascent stage of neural network algorithms and applications, and the underdevelopment of advanced optical components essential for ONNs, resulted in sluggish progress in ONN research for a considerable period. Until recently, the emergence of massive data and the flourishing development of machine learning algorithms such as deep learning have provided beneficial conditions for advancing neural networks, making their applications in various industries more extensive and in-depth. At present, the artificial intelligence era is rising. Nonetheless, electronic computing hardware is encountering physical structure limitations that hinder further enhancements, leading to a predicament where it cannot furnish the greater computing resource support required by the burgeoning intelligent era. Thus, there is an urgent need to find a new computing paradigm to provide greater computing resource support. ONNs represent such a paradigm, offering numerous advantages over traditional electronic hardware computing architectures. For instance, ONNs exhibit exceptionally low energy consumption during inference calculations and benefit from their capacity to support high-speed parallel signal processing, resulting in faster processing speeds than their electronic counterparts. It is crucial to highlight that the enhancement of ONNs in task processing speed is not solely reliant on the miniaturization of device structures. Furthermore, the current state of semiconductor processing technology can fully meet the fabrication requirements for optical devices. Therefore, ONNs have received extensive attention and progress in recent years.

ONNs have several advantages over electronic neural networks, including low latency, low power consumption, and ultra-fast parallel signal processing. Neural networks need to process large matrix operations when performing inference tasks repeatedly, and the propagation of light is a natural matrix operation process. Consequently, light possesses an intrinsic advantage over traditional electronic hardware in executing large-scale linear matrix operations. The architecture of ONNs represents a novel computing paradigm that holds the potential to overcome the impending bottleneck in the upgrading of computing resources. Current research on ONNs primarily explores two domains: spatial bulk ONNs and silicon-based integrated ONNs. Among them, spatial bulk ONNs include free-space diffractive ONNs and other spatial bulk ONNs; silicon-based integrated ONNs include integrated interference ONNs, integrated wavelength-division ONNs, integrated diffractive ONNs, and other integrated ONNs. In the subsequent section, this thesis will provide a detailed overview of the research status of the ONNs mentioned above, analyze the improvement space of current ONNs research work, and propose the research content, significance, and objectives of this thesis.

1.2 Research on Non-integrated ONNs

This section is structured into three parts to elucidate the current research status of non-integrated ONNs, encompassing free-space diffractive ONNs, lens-based ONNs, and other spatial bulk ONNs.

1.2.1 Non-integrated Free-Space Diffractive ONNs

Non-integrated free-space diffractive ONNs are primarily composed of individual optical components such as digital micromirror devices (DMD) [17], spatial light modulators (SLM) [17–20], and optical diffraction plates [21–29].

In 2018, Lin et al. [21] proposed a free-space diffractive optical neural network (DONN) based on diffraction theory and 3D printing technology. The team utilized the Rayleigh-Sommerfeld diffraction equation to express the propagation of the spatial light field analytically and pre-trained the structure parameters of the free-space DONN on a computer. Then, they adopted 3D printing technology to implement these parameters physically and subsequently completed the inference function of the free-space DONN in a physical hardware setup. The DONN achieved a 91.75% accuracy rate of blind testing on the MNIST dataset, and the inference calculation process was passive, significantly reducing system power consumption. The free-space DONN was achieved by designing the thickness of the diffractive structures at different positions on the diffraction plate to regulate the phase or amplitude changes of the light so that the incident light would produce interference as it propagated, thereby realizing the neural network's inference function. This implementation of the free-space DONN was non-reconfigurable and could only perform specific inference functions with a working wavelength of 0.75 mm. In 2019, Wu et al. [30] developed a metasurface operating in the visible light spectrum (at a wavelength of 700 nm) composed of rectangular prisms. The height of the rectangular prisms was fixed at 600 nm, while the spacing between adjacent rectangular prisms (i.e., the arrangement period) was fixed at 235 nm. The width of the rectangular prism varied from 50 to 180 nm, and the team achieved phase and amplitude control of each diffraction unit by changing the width of the rectangular prisms. This study conducted numerical simulations of free-space DONNs and theoretically validated their performance using the MNIST dataset. Additionally, the authors investigated the angle of incident light as it traveled to the metasurface and provided the phase and amplitude modulation of the unit cells under different incident angles. This work provides a comprehensive discussion of the incident angle, which is meaningful in guiding similar future works. In 2020, Qian et al. [22] engineered a free-space DONN based on diffraction theory and mechanical processing technology that could perform optical logical operations. They demonstrated logical operations such as AND, OR, and NOT in an optical manner. The free-space DONN made by 3D printing or mechanical processing technologies, as in this work, has a larger working wavelength because a higher diffraction efficiency often requires the diffractive units to be of the same size as the working wavelength, and a larger working wavelength also requires a larger diffractive structure, making it easier to fabricate and process, such as the working wavelength of 17.6 mm in this work. In 2021, Zhou et al. [17] advanced significantly by developing a reconfigurable free-space DONN utilizing DMD and SLM. This innovation endowed free-space DONNs with reconfigurability and enhanced generality, broadening their applicability and computing efficiency. Despite these advancements, this work, along with similar studies, still encountered the following challenges: firstly, the physical bulk of such free-space DONNs often results in systems that are cumbersome and unstable, limiting their practicality; secondly, the alignment and calibration process between the individual optical components may introduce additional errors to the entire system.

The previously discussed free-space DONNs were composed of discrete optical components, and the calibration process between the individual components could lead to additional system errors, resulting in poor stability and portability. To address the calibration error problem between the individual components, Goi et al. [26] developed a highly integrated free-space DONN using double-photon nanolithography technology based on a complementary metal–oxide–semiconductor (CMOS) chip substrate. This approach involved creating multiple layers of diffractive metasurfaces, composed of subwavelength cylindrical arrays with a height variation step of 10 nm. This high level of integration facilitated the creation of approximately 500 million neurons per square centimeter, operating at a wavelength of 785 nm. In addition, Luo et al. [28] in 2022 engineered a multifunctional DONN consisting of metasurfaces based on a CMOS chip substrate, which could perform multi-channel sensing and multi-task processing in a visible light environment. This metasurface was composed of an array of nanopillars with subwavelength structures, where the height of the nanopillars was fixed at 600 nm, and the center spacing (i.e., the period) between adjacent nanopillars was 400 nm. The phase of propagating light could be modulated by changing the length and width of each nanopillar. This work utilized the polarization multiplexing scheme of subwavelength nanopillar structures to construct a multi-channel classifier framework. It achieved classification recognition for different tasks of MNIST and Fashion-MNIST. The integration density was approximately 6.25×10^6 neurons per square millimeter, and the operating wavelength was 532 nm. While the fabrication of metasurfaces with subwavelength structures on CMOS substrates for constructing free-space DONNs resolved the alignment and calibration issues associated with discrete optical components, it introduced significant processing challenges and costs. Moreover, achieving reconfigurability in such subwavelength structures across various metasurfaces remains a formidable challenge.

1.2.2 Non-integrated Lens-Based ONNs

Non-integrated lens-based ONNs are typically built up by a 4*f* lens group system [31–33]. In 2018, Chang et al. [31] used DMD as the signal input mechanism, placed a phase mask template in the Fourier plane of a 4*f* system (an optical system consisting of a set of optical lenses), and achieved the convolution operation function by modulating the Fourier plane spectrum. This implementation of the convolution function was characterized by near-zero energy consumption. When the camera captured the spectrum-modulated relevant information, it processed this data and executed the nonlinear operations in the electrical domain. This approach created a photoelectric hybrid neural network, or an electrically assisted ONN, where the roles of light and electricity were distinctly optimized: light handled the extensive linear matrix operations, while electricity managed the nonlinear calculations, which involved a relatively minor computational load. In 2019, Yan et al. [32] used the refractive properties of crystals to modulate image information in the Fourier plane or image plane to achieve deep free-space ONNs. This work introduced nonlinear correlation parameters of ferroelectric thin films [34, 35] into the numerical calculation process and introduced optical nonlinear functions at the end of each hidden layer during training. The performance of the free-space ONNs was compared and analyzed based on the training results with or without ferroelectric thin film nonlinear layers. In 2020, Miscuglio et al. [33] developed a large-scale parallel signal processing pure-amplitude free-space Fourier ONNs based on lens groups and DMDs, in which DMDs achieved the input and Fourier plane amplitude modulation. Therefore, the free-space Fourier ONNs based on the 4*f* system had reconfigurable capabilities, enhancing their computational versatility and broadening their potential applications.

1.2.3 Other Non-integrated Spatial Bulk ONNs

Other non-integrated spatial bulk ONNs [36–38] refer to neural networks composed of discrete optical components such as standard single-mode fibers, dispersion fibers, modulators, and band-pass filters. In 2021, Xu et al. [36] developed a photonic convolution accelerator utilizing the integrated optical frequency comb, the bulk electro-optic modulator (EOM), the standard single-mode optical fiber, and the photodetector (PD). Specifically, the EOM modulated the intensity at predetermined time slots across different wavelengths, and the PD detected the aggregate power of these wavelengths at specific times, thereby functionally realizing the equivalent operation of a convolution kernel. The operating speed of the photonic convolution kernel could reach tens of trillions of operations per second (11 TOPS), and the spatial bulk ONNs achieved an accuracy of 88% on the MNIST dataset in a blind test. In the same year, Stelzer et al. [37] constructed the computational function of a single neuron using a nonlinear device and multiple time-delay feedback loops. They engineered an arbitrary-sized deep neural network by continuously reconstructing the parameters of the single neuron.

1.3 Research on Integrated ONNs

In Sect. 1.2, the non-integrated ONNs were introduced, which can be constructed with discrete optical components such as traditional lenses, DMDs, SLMs, and two-dimensional optical diffraction plates, or with diffraction plates on CMOS chip substrates, or with other discrete optical components such as dispersive optical fibers, modulators, and band-pass filters. Among these, discrete optical components have the disadvantages of large size, poor stability, and poor portability, which may introduce additional errors in the ONN system. The ONNs fabricated on CMOS chip substrates overcome the alignment and calibration issues between discrete components. However, the fabrication is difficult and costly, and it is challenging to realize subwavelength structures with reconfigurability between diffractive layers. Both scenarios limit the application of such kinds of free-space ONNs. Silicon-based integrated ONNs designed on a silicon-on-insulator (SOI) platform are fabricated through CMOS-compatible processes, enabling cost-effective mass production via advanced photolithography and exposure techniques. This approach can simultaneously address the issues encountered by non-integrated spatial bulk ONNs. This section will introduce silicon-based integrated interferometric ONNs [39–47], integrated wavelength-division ONNs [48–59], integrated diffractive ONNs [60–63], and other silicon-based integrated ONNs [64–69].

1.3.1 Silicon-Based Integrated Interferometric ONNs

In 2017, Shen et al. [39] proposed an on-chip interferometric ONN based on a SOI platform. The team utilized 56 programmable MZIs to complete the optical matrix multiplication in a topological cascading manner, thus realizing the matrix operations of the neural network. The specific operation was to decompose any real-valued matrix $\boldsymbol{M}$ into a diagonal matrix $\boldsymbol{\Sigma}$ and two unitary matrices $\boldsymbol{U}$ and $\boldsymbol{V}$ through the singular value decomposition method, i.e., $\boldsymbol{M} = \boldsymbol{U\Sigma V}^{\dagger}$, where the symbol '†' denoted the conjugate operation of matrix $\boldsymbol{V}$. Theoretical analysis showed that $\boldsymbol{U}$ and $\boldsymbol{V}^{\dagger}$ in any unitary matrix transformation could be realized by optical beam splitters and phase shifters. The diagonal matrix $\boldsymbol{\Sigma}$ could be realized by optical attenuators. These functions could be realized in physical devices through the topological cascading of MZIs. In 2018, Hughes et al. [40] proposed an in-situ training method based on cascading MZIs to optimize the structure parameters of ONNs, making the training of ONNs more efficient. In 2021, Zhang et al. [44] designed a complex-valued ONN based on the MZI mesh. This work simultaneously utilized the phase and amplitude of light, thereby increasing the number of adjustable variables in the network.

This design method improved the performance of ONNs based on MZIs, significantly enhancing the ONN's computing speed and energy efficiency. Additionally, the learning capabilities of the complex-valued neural network, including classification accuracy and the convergence speed of the loss function, were superior to those of real-valued neural networks. In this study, the complex-valued neural network was used to complete tasks such as logic gate operations, IRIS dataset category prediction, nonlinear data (circle and spiral) classification, and MNIST dataset category prediction. The following year, Zhu et al. [46] introduced diffraction units (implementing the function of the Fourier transform and the inverse Fourier transform) into the MZI mesh. Regarding the same MNIST and Fashion-MNIST dataset classification experiments, combining MZI mesh and diffraction units reduced the area and energy consumption of the entire ONNs to 1/10 of that of ONNs implemented with only traditional MZIs.

1.3.2 Silicon-Based Integrated WDM-System ONNs

In 2019, Feldmann et al. [51] adopted micro-ring wavelength division multiplexing (WDM) systems and PCM materials based on the Si_3N_4 platform to design on-chip optical devices that imitate the functions of neurons and synapses. When these devices were constructed into a network or neuromorphic system, they processed information in a way that was more like the human brain. During the implementation of these ONNs, the input pulses were weighted through PCM units. Then, the corresponding wavelength of light was coupled into a single-mode waveguide through the micro-ring for power addition. When the integrated power in the single-mode waveguide exceeded a certain threshold, the PCM unit on the last ring resonator switched and generated an output pulse, thus completing a reasoning calculation on the optical domain. During this calculation, the nonlinear function was directly realized optically. In 2021, Feldmann et al. [55] implemented an integrated photonic tensor core on the Si_3N_4 platform based on micro-ring WDM systems and PCM, which realized the parallel processing function of traditional convolutional kernels on the chip. It could run at a speed of trillions of operations per second (i.e., 10^{12} multiply-accumulate operations per second). This work utilized PCM material storage arrays and optical frequency combs based on the photonic chip to realize parallel photonic memory computing, which could operate at a bandwidth of over 14 GHz, and its computing speed was limited only by the modulator and photoelectric detector speed. In the same year, Huang et al. [56] designed a WDM-system ONN based on an SOI platform, which was used to compensate for the nonlinear problems in submarine optical fiber transmission systems. The reconfigurable WDM-system ONN could process optical signals in the analog domain, significantly alleviating the stringent requirements of traditional digital signal processing circuits in terms of complexity and speed. In 2023, Bai et al. [59] designed a photonic processing unit using an on-chip micro-ring WDM system, which included multiple wavelength sources, data loading areas, and data processing centers. The photonic processing unit could use the time-wavelength plane stretching method to implement the convolution function, with a computing density of approximately 1.04 TOPS per square millimeter (i.e., approximately 1 trillion operations per square millimeter per second), which was five times the highest computing density of any silicon-based integrated WDM-system ONNs before. The convolutional neural network designed based on the photonic processing unit achieved image edge detection quality and MNIST data recognition accuracy (96.6%) which was comparable to the detection and recognition results of its electrical counterparts.

1.3.3 Silicon-Based Integrated Diffractive ONNs

Silicon-based integrated ONNs implemented based on the cascading MZI mesh or micro-ring WDM system have overcome issues such as the large volume, poor portability and stability of free-space DONNs, and possible system errors caused by the alignment and calibration process of discrete optical components. However, both silicon-based integrated ONNs realized by MZI topology cascading and those realized by micro-ring WDM systems have a common problem: low computing capacity. To solve this problem, research on silicon-based integrated DONNs has been reported one after another [60–62]. In 2020, Zarei et al. [60] designed a silicon-based integrated DONN with subwavelength rectangular micro-slot structures and verified its performance in a simulation manner. The silicon-based integrated DONN had a very high integration degree, with a theoretical calculation of approximately 3.75 $\times$ 10^4 neurons per square millimeter, and its computing capacity was about 1.54 $\times$ 10^{16} operations per square millimeter per second, i.e., 1.54 $\times$ 10^4 TOPS. In 2022, Zarei et al. [62] used the silicon-based integrated DONN to realize the logical gate operation function. In the same year, Wang et al. [61] experimentally realized the silicon-based integrated DONN designed with subwavelength rectangular micro-slot arrays (metalines). The area of the silicon-based integrated DONN was about 0.135 mm^2, and the theoretical calculation showed that approximately 6.7 $\times$ 10^3 neurons could be integrated per square millimeter. Before the research on silicon-based integrated DONNs began, some studies had already been related to silicon-based on-chip diffraction, such as on-chip lenses [70, 71].

1.3.4 Other Integrated ONNs

Other integrated ONNs [64–68] refer to neural networks constructed with two-dimensional silicon-based integrated optical waveguides, on-chip multi-mode interferometers (MMIs), phase shifters or attenuators, and three-dimensional integrated optical waveguides and other optical components. In 2020, Moughames et al. [65]

used two-photon polymer printing technology to fabricate a three-dimensional integrated low-loss photonic waveguide array, where the diameter of the photonic waveguide was 1.2 μm, the spacing between adjacent waveguides was 20 μm, and approximately 2200 neurons could be integrated per cubic millimeter. In the same year, Qu et al. [66] used reverse engineering to design and fabricate silicon-based nanophotonic scattering units. The optical scattering units were fabricated within 4×4 μm through reverse design. This was achieved by fabricating nanopatterns with high degrees of freedom in the MMI region. Theoretically, these optical scattering units could also perform arbitrary unitary matrix multiplication. Based on this result, the team designed a silicon-based integrated ONN for the MNIST dataset and achieved a blind test accuracy of 97.1% on the MNIST testing set. In 2022, Ashtiani et al. [68] adopted carrier doping technology to fabricate electrically controlled tunable attenuators on silicon waveguides. Thus, the weight configuration of the neural network could be realized by adjusting the attenuation coefficient of each waveguide. The optical power output from each waveguide would be collected separately by multiple PDs and then it would be output through a micro-ring modulator designed based on carrier injection. This process could be used to realize optical nonlinear functions. The silicon-based integrated system could only perform a single neuron operation simultaneously. To complete a matrix operation, it was necessary to adjust the attenuator coefficients multiple times.

1.4 Research Significance and Objectives

In summary, the lithography precision of semiconductor manufacturing processes has approached the physical limit of transistors. Meanwhile, with the rapid development of artificial intelligence, neural network algorithms will require more computing resources and energy consumption. However, the current electronic computing hardware will be unable to meet future computing demands. ONNs are a potential solution to this problem, but current free-space DONNs are composed of discrete optical components, so they are large, have poor stability, and have poor portability. The fabrication of integrated ONNs based on the CMOS substrate is highly challenging, the cost of mass production is prohibitively expensive, and the reconfigurability of computing units on the interlayer diffraction plates is also extremely difficult to achieve. For silicon-based integrated ONNs other than integrated DONNs, the computing units need continuous energy supply during operation. Additionally, the relatively large size of these computing units poses a challenge to large-scale expansion, resulting in the generally low computing capacity of this type of silicon-based integrated ONNs. However, in the past three years, for silicon-based integrated DONNs, apart from the experimental work by Wang et al. [61] in 2022, other studies have been confined to the theoretical calculation stage [60, 62, 72]. In addition, in the work proposed by Wang et al. [61], signal loading is achieved through the combination of lens groups and DMD, which has not been realized in a silicon-integrated manner. Furthermore, there has been no in-depth exploration of the weight mapping

Table 1.1 Research objectives of this thesis

Parameters	Dimension	Size of the computing unit	Theoretical integration degree (computing units/ mm^2)	Energy efficiency	Stability/ portability	Computing speed (operations/ s)
Objective	Integrated	$\leq$ 1.5μm × 2μm	~10^3	High	Good	>10^{12}

model for the on-chip structure parameters of silicon-based integrated DONN. Also, there has been no mention of the compensation scheme and related measures for the experimental system error during the experiment. At present, there is no comprehensive research on silicon-based integrated DONNs, such as the forward propagation model of silicon-based integrated DONNs, the weight mapping rules for the structure parameters of integrated DONNs, and the practical problems that may be encountered in experimental fabrication, packaging, and testing of the integrated DONN chips.

Therefore, this thesis will conduct a systematic study on aspects such as theoretical exploration, structure design, and simulation calculation of silicon-based integrated DONNs, as well as the fabrication, packaging, and testing aspects of silicon-based integrated DONN chips and compensation for experimental system errors. First, a silicon-based integrated DONN model will be built during the research process, including an on-chip forward propagation model, a silicon-based integrated DONN training model, and a weight mapping model. Second, corresponding silicon-based integrated DONN chips will be designed for various tasks, and their performance will be successfully verified through simulation and experiment. Additionally, since system errors are unavoidable in chip fabrication and experimental testing processes, this work will analyze system errors in detail and propose effective and feasible error compensation schemes. Finally, to further improve the performance of silicon-based integrated DONN chips, a design method for the multi-functional silicon-based integrated DONN system architecture will be proposed. The research objectives are summarized in Table 1.1, and the research results will provide valuable references for the design and experimental work of silicon-based integrated DONN chips.

1.5 Main Content and Structure of the Thesis

This research aims to systematically study key scientific and technological issues related to silicon-based integrated DONNs, including the forward propagation model of the silicon-based integrated DONNs, namely the two-dimensional spatial domain electromagnetic propagation model (SDEPM), used to represent the propagation laws of light in silicon-based slab waveguides. The second is the phase weight mapping model of the silicon-based integrated DONNs, which ensures that the phase weight parameters obtained by pre-training on a computer can be accurately implemented on the physical structure. Then, based on the silicon-based integrated DONN model, comprehensive research is conducted on the structure design, simulation verification, chip fabrication, packaging, testing, and system error compensation of the silicon-based integrated DONN chips.

This thesis consists of five chapters, and the main contents of each chapter are arranged as follows:

Chapter 1 will introduce the impact of semiconductor process development on computing resources and how the abundance of computing resources has driven the widespread application and remarkable achievements of neural networks in various industries. However, as the processing accuracy of semiconductor technology is approaching the physical limit of transistors, further improvements in the computing resources of traditional computing hardware will face a bottleneck. At this point, a new computing paradigm is needed to break through the bottleneck in further improving computing resources. Unlike electronic computing architecture, the computing process of ONNs is based on the interference and diffraction of light, and the entire computing process is passive, with advantages such as low latency, low power consumption, and large-scale parallel signal processing. Therefore, this chapter will deeply introduce the current research status of ONNs. From the current research status of ONNs, it will be found that there are still many limitations in related studies, so further objectives and significance of the research work in this thesis will be proposed. Finally, the content arrangement of each chapter of this thesis will be introduced.

Chapter 2 will introduce the research on silicon-based integrated DONN model, including the two-dimensional silicon-based SDEPM, the training model, and the phase weight mapping mode. Among them, the light propagation law of the silicon-based integrated slab waveguide is inspired by the Huygens-Fresnel principle under a specific situation and then compared with the 2.5D FDTD optical field evolution to obtain the two-dimensional silicon-based SDEPM. For the phase weight mapping model, the effective refractive index of the subwavelength rectangular micro-slot region will be calculated respectively using the effective medium theory and the 2.5D FDTD method. Furthermore, the stability of the effective refractive index of the subwavelength rectangular micro-slot region will be verified under light sources with different incident angles. Finally, the phase weight mapping model's reliability, stability, and effectiveness are ensured under certain approximate conditions.

Chapter 3 will introduce the specific silicon-integrated DONN chips designed based on the proposed DONN model for binary, ternary, and ten classification tasks. The performance of the silicon-integrated DONN chip for binary classification will be verified by means of numerical computation and 2.5D FDTD method. The performance of the ternary classification and ten classification silicon-integrated DONN chips will be experimentally verified. This chapter will systematically introduce the theoretical analysis, structure design, simulation verification, layout drawing, chip fabrication, packaging, testing, and experimental system error compensation scheme of the silicon-integrated DONN chips. During the experimental process, some practical issues that do not need to be considered in the simulation process will be involved, such as the attenuation structure used in the output interface and the mode selection at the output stage.

Chapter 4 will propose a design method for a multi-functional silicon-based integrated DONN system architecture to address the limitations of dedicated silicon-based integrated DONN chips. According to this design method, the silicon-based integrated DONN chips can process high-speed serial input signals and handle multiple tasks. This chapter will validate the performance of the multi-functional silicon-based integrated DONN chip in two parts. Firstly, the feasibility of converting high-speed serial input signals to parallel outputs using arrayed waveguide serial-to-parallel conversion technology will be verified experimentally. Secondly, multi-task learning and transfer learning methods will be used to design a silicon-based integrated DONN chip with multi-task processing capability, and its performance will be verified through simulation calculations.

Chapter 5 will summarize the main research achievements and innovations of the thesis and provide an outlook on the aspects that can be further explored in the future regarding this research work.

References

1. Liu YQ, Qian K, Wang K et al (2022) Effective scaling of blockchain beyond consensus innovations and Moore's law: Challenges and opportunities. IEEE Syst J 16(1):1424–1435
2. LeCun Y, Bengio Y, Hinton G (2015) Deep learning. Nature 521(7553):436–444
3. Samarasinghe S (2016) Neural networks for applied sciences and engineering: from fundamentals to complex pattern recognition. Auerbach publications
4. Abiodun OI, Jantan A, Omolara AE et al (2018) State-of-the-art in artificial neural network applications: a survey. Heliyon 4(11)
5. Gu J, Wang Z, Kuen J et al (2018) Recent advances in convolutional neural networks. Pattern Recogn 77:354–377
6. Parisi GI, Kemker R, Part JL et al (2019) Continual lifelong learning with neural networks: a review. Neural Netw 113:54–71
7. Samek W, Montavon G, Lapuschkin S et al (2021) Explaining deep neural networks and beyond: a review of methods and applications. Proc IEEE 109(3):247–278
8. Miikkulainen R, Liang J, Meyerson E et al (2024) Evolving deep neural networks. Artificial intelligence in the age of neural networks and brain computing. Academic Press, pp 269–287
9. Krizhevsky A, Sutskever I, Hinton GE (2017) ImageNet classification with deep convolutional neural networks. Commun ACM 60(6):84–90
10. Ananthanarayana T, Srivastava P, Chintha A et al (2021) Deep learning methods for sign language translation. ACM Trans Access Comput (TACCESS) 14(4):1–30
11. Graves A, Mohamed A, Hinton G (2013) Speech recognition with deep recurrent neural networks. In: 2013 IEEE international conference on acoustics, speech and signal processing. IEEE, pp 6645–6649
12. Rausch V, Hansen A, Solowjow E et al (2017) Learning a deep neural net policy for end-to-end control of autonomous vehicles. In: 2017 American control conference (ACC). IEEE, pp 4914–4919
13. Litjens G, Kooi T, Bejnordi BE et al (2017) A survey on deep learning in medical image analysis. Med Image Anal 42:60–88
14. Baldi P (2018) Deep learning in biomedical data science. Annu Rev Biomed Data Sci 1(1):181–205

15. Apostolidis KD, Papakostas GA (2021) A survey on adversarial deep learning robustness in medical image analysis. Electronics 10(17):2132
16. Caulfield HJ, Kinser J, Rogers SK (1989) Optical neural networks. Proc IEEE 77(10):1573–1583
17. Zhou T, Lin X, Wu J et al (2021) Large-scale neuromorphic optoelectronic computing with a reconfigurable diffractive processing unit. Nat Photonics 15(5):367–373
18. Zuo Y, Li B, Zhao Y et al (2019) All-optical neural network with nonlinear activation functions. Optica 6(9):1132–1137
19. Zhou T, Fang L, Yan T et al (2020) In situ optical backpropagation training of diffractive optical neural networks. Photonics Res 8(6):940–953
20. Wang T, Ma SY, Wright LG et al (2022) An optical neural network using less than 1 photon per multiplication. Nat Commun 13(1):123
21. Lin X, Rivenson Y, Yardimci NT et al (2018) All-optical machine learning using diffractive deep neural networks. Science 361(6406):1004–1008
22. Qian C, Lin X, Lin X et al (2020) Performing optical logic operations by a diffractive neural network. Light: Sci Appl 9(1):59
23. Dou H, Deng Y, Yan T et al (2020) Residual D2NN: training diffractive deep neural networks via learnable light shortcuts. Opt Lett 45(10):2688–2691
24. Fang T, Li J, Zhang X et al (2021) Classification accuracy improvement of the optical diffractive deep neural network by employing a knowledge distillation and stochastic gradient descent β-Lasso joint training framework. Opt Express 29(26):44264–44274
25. Wang P, Xiong W, Huang Z et al (2021) Orbital angular momentum mode logical operation using optical diffractive neural network. Photonics Res 9(10):2116–2124
26. Goi E, Chen X, Zhang Q et al (2021) Nanoprinted high-neuron-density optical linear perceptrons performing near-infrared inference on a CMOS chip. Light: Sci Appl 10(1):40
27. Shi W, Huang Z, Huang H et al (2022) LOEN: lensless opto-electronic neural network empowered machine vision. Light: Sci App 11(1):121
28. Luo X, Hu Y, Ou X et al (2022) Metasurface-enabled on-chip multiplexed diffractive neural networks in the visible. Light: Sci App 11(1):158
29. Sun Y, Dong M, Yu M et al (2021) Modeling and simulation of all-optical diffractive neural network based on nonlinear optical materials. Opt Lett 47(1):126–129
30. Wu Z, Zhou M, Khoram E et al (2019) Neuromorphic metasurface. Photo Res 8(1):46–50
31. Chang J, Sitzmann V, Dun X et al (2018) Hybrid optical-electronic convolutional neural networks with optimized diffractive optics for image classification. Sci Rep 8(1):1–10
32. Yan T, Wu J, Zhou T et al (2019) Fourier-space diffractive deep neural network. Phys Rev Lett 123(2):023901
33. Miscuglio M, Hu Z, Li S et al (2020) Massively parallel amplitude-only Fourier neural network. Optica 7(12):1812–1819
34. Christodoulides DN, Coskun TH, Mitchell M et al (1997) Theory of incoherent self-focusing in biased photorefractive media. Phys Rev Lett 78(4):646
35. Waller L, Situ G, Fleischer JW (2012) Phase-space measurement and coherence synthesis of optical beams. Nat Photonics 6(7):474–479
36. Xu X, Tan M, Corcoran B et al (2021) 11 TOPS photonic convolutional accelerator for optical neural networks. Nature 589(7840):44–51
37. Stelzer F, Röhm A, Vicente R et al (2021) Deep neural networks using a single neuron: folded-in-time architecture using feedback-modulated delay loops. Nat Commun 12(1):5164
38. Zhang L, Li C, He J et al (2021) Optical machine learning using time-lens deep neural networks Photonics. MDPI 8(3):78
39. Shen Y, Harris NC, Skirlo S et al (2017) Deep learning with coherent nanophotonic circuits. Nat Photonics 11(7):441–446
40. Hughes TW, Minkov M, Shi Y et al (2018) Training of photonic neural networks through in situ backpropagation and gradient measurement. Optica 5(7):864–871
41. Williamson IAD, Hughes TW, Minkov M et al (2019) Reprogrammable electro-optic nonlinear activation functions for optical neural networks. IEEE J Sel Top Quantum Electron 26(1):1–12

42. Zhang T, Wang J, Dan Y et al (2019) Efficient training and design of photonic neural network through neuroevolution. Opt Express 27(26):37150–37163
43. Deng H, Khajavikhan M (2021) Parity–time symmetric optical neural networks. Optica 8(10):1328–1333
44. Zhang H, Gu M, Jiang XD et al (2021) An optical neural chip for implementing complex-valued neural network. Nat Commun 12(1):457
45. Shi Y, Ren J, Chen G et al (2022) Nonlinear germanium-silicon photodiode for activation and monitoring in photonic neuromorphic networks. Nat Commun 13(1):6048
46. Zhu HH, Zou J, Zhang H et al (2022) Space-efficient optical computing with an integrated chip diffractive neural network. Nat Commun 13(1):1044
47. Wang X, Xie P, Chen B et al (2022) Chip-based high-dimensional optical neural network. Nano-Micro Letters 14(1):221
48. Tait AN, Nahmias MA, Shastri BJ et al (2014) Broadcast and weight: an integrated network for scalable photonic spike processing. J Lightwave Technol 32(21):3427–3439
49. Tait AN, De Lima TF, Zhou E et al (2017) Neuromorphic photonic networks using silicon photonic weight banks. Sci Rep 7(1):7430
50. Tait AN, Wu AX, Ferreira de Lima T et al (2018) Two-pole microring weight banks. Optics Letters 43(10):2276–2279
51. Feldmann J, Youngblood N, Wright CD et al (2019) All-optical spiking neurosynaptic networks with self-learning capabilities. Nature 569(7755):208–214
52. Wright CD, Bhaskaran H, Pernice WHP (2019) Integrated phase-change photonic devices and systems. MRS Bull 44(9):721–727
53. Tait AN, Ferreira de Lima T, Nahmias MA et al (2019) Silicon photonic modulator neuron. Phys Rev App 11(6):064043
54. Ohno S, Toprasertpong K, Takagi S et al (2020) Si microring resonator crossbar arrays for deep learning accelerator. Japanese J App Phys 59(SG):SGGE04
55. Feldmann J, Youngblood N, Karpov M et al (2021) Parallel convolutional processing using an integrated photonic tensor core. Nature 589(7840):52–58
56. Huang C, Fujisawa S, de Lima TF et al (2021) A silicon photonic–electronic neural network for fibre nonlinearity compensation. Nat Elect 4(11):837–844
57. Sludds A, Bandyopadhyay S, Chen Z et al (2022) Delocalized photonic deep learning on the internet's edge. Science 378(6617):270–276
58. Xu S, Wang J, Yi S et al (2022) High-order tensor flow processing using integrated photonic circuits. Nat Commun 13(1):7970
59. Bai B, Yang Q, Shu H et al (2023) Microcomb-based integrated photonic processing unit. Nat Commun 14(1):66
60. Zarei S, Marzban M, Khavasi A (2020) Integrated photonic neural network based on silicon metalines. Opt Express 28(24):36668–36684
61. Wang Z, Chang L, Wang F et al (2022) Integrated photonic metasystem for image classifications at telecommunication wavelength. Nat Commun 13(1):2131
62. Zarei S, Khavasi A (2022) Realization of optical logic gates using on-chip diffractive optical neural networks. Sci Rep 12(1):15747
63. Yan T, Yang R, Zheng Z et al (2022) All-optical graph representation learning using integrated diffractive photonic computing units. Sci Adv 8(24):eabn7630
64. Hughes TW, Williamson IAD, Minkov M et al (2019) Wave physics as an analog recurrent neural network. Sci Adv 5(12):eaay6946
65. Moughames J, Porte X, Thiel M et al (2020) Three-dimensional waveguide interconnects for scalable integration of photonic neural networks. Optica 7(6):640–646
66. Qu Y, Zhu H, Shen Y et al (2020) Inverse design of an integrated-nanophotonics optical neural network. Sci Bull 65(14):1177–1183
67. Zhao X, Lv H, Chen C et al (2021) On-chip reconfigurable optical neural networks
68. Ashtiani F, Geers AJ, Aflatouni F (2022) An on-chip photonic deep neural network for image classification. Nature 606(7914):501–506

69. Huang Y, Wang W, Qiao L et al (2022) Programmable low-threshold optical nonlinear activation functions for photonic neural networks. Opt Lett 47(7):1810–1813
70. Zhang J, Yang J, Xin H et al (2017) Ultrashort and efficient adiabatic waveguide taper based on thin flat focusing lenses. Opt Express 25(17):19894–19903
71. Wang Z, Li T, Soman A et al (2019) On-chip wavefront shaping with dielectric metasurface. Nat Commun 10(1):3547
72. Fu T, Zang Y, Huang H et al (2021) On-chip photonic diffractive optical neural network based on a spatial domain electromagnetic propagation model. Opt Express 29(20):31924–31940

Chapter 2
Silicon-Based Integrated Diffractive Optical Neural Network Model

Abstract This chapter proposes a silicon-based integrated DONN model, which mainly consists of three parts: the two-dimensional silicon-based SDEPM, the training model of the silicon-based integrated DONNs, and the phase weight mapping model. Among these, the two-dimensional silicon-based SDEPM serves as the foundation for the training model of the silicon-based integrated DONN. This is because the two-dimensional silicon-based SDEPM can analytically describe the propagation process of light in slab waveguides based on an SOI platform. With the analytical expression, it is possible to pre-train the key structure parameters of the silicon-based integrated DONN using algorithms such as gradient descent on a computer. Once the silicon-based integrated DONN key structure parameters are obtained, the phase weight mapping model can accurately map the parameters onto the silicon-based integrated DONN chip. Specifically, Sect. 2.1 will introduce the two-dimensional silicon-based SDEPM. Section 2.2 will introduce the training model of the silicon-based integrated DONN. Section 2.3 will discuss the phase weight mapping model. Section 2.4 serves as the summary of this chapter, providing an analysis and summary of each part of the proposed models in this chapter.

2.1 Two-Dimensional Spatial Domain Electromagnetic Propagation Model

This section presents the two-dimensional silicon-based SDEPM. Inspired by the Huygens-Fresnel principle in free space, this advanced model incorporates phase and amplitude correction terms, enabling accurate analytical characterization of light propagation in silicon-based slab waveguides. The model calibration uses 2.5D FDTD simulation results as the reference benchmark. Based on this SDEPM, a computational model for forward propagation in silicon-based integrated DONNs can be effectively implemented on a computer. The two-dimensional silicon-based SDEPM serves as the basis for the design of silicon-based integrated DONNs. Leveraging this foundation, the structure parameters of the silicon-based integrated DONN

© Tsinghua University Press 2026

T. Fu, *Research on Silicon-Based Integrated Diffractive Optical Neural Network*, Springer Theses, https://doi.org/10.1007/978-981-95-3787-7_2

can be iteratively optimized through forward propagation, gradient descent, and error backpropagation algorithms.

This subsection begins by expounding on the Huygens-Fresnel principle, the fundamental theory of wave optics. This principle, formulated through the combined contributions of Christiaan Huygens and Augustin-Jean Fresnel, provides a comprehensive framework for understanding wave propagation and diffraction phenomena. The Huygens principle can be used to describe the propagation process of light waves, as shown in Fig. 2.1a, where there are two wavefronts Σ and Σ'. Among them, Σ represents the phase distribution of the light source S at a certain moment, and each point on Σ can be regarded as a secondary wave source. After these secondary wave sources propagate for a certain period, the locus of their wavefronts forms a new wavefront Σ' at a later moment. During this process, the light wave propagates along the normal direction of the wavefront.

Later, Fresnel supplemented Huygens' principle with the interference theory, leading to the Huygens-Fresnel principle. According to this principle, as shown in Fig. 2.1b, the effect of a light source S on any point P after a particular propagation time can be regarded as the superposition of light waves emitted by all secondary wave sources on an arbitrary wavefront between S and P. At this time, the complex amplitude $\tilde{\boldsymbol{E}}_{\mathrm{Q}}$ at any point Q on the wavefront Σ (ZZ') is given by $\tilde{\boldsymbol{E}}_{\mathrm{Q}} = Ae^{jkR}/R$, where A is the amplitude at a unit distance from the monochromatic point source S, and R is the radius of the wavefront Σ. If an elemental wavefront $d\sigma$ is taken at point Q, the complex amplitude produced at point P by the secondary wave source on $d\sigma$ can be calculated, as shown in Eq. 2.1.

$$\mathrm{d}\widetilde{E}(P) = CK(\theta)\widetilde{E}_Q \cdot \frac{e^{jk_0 r}}{r} d\sigma \tag{2.1}$$

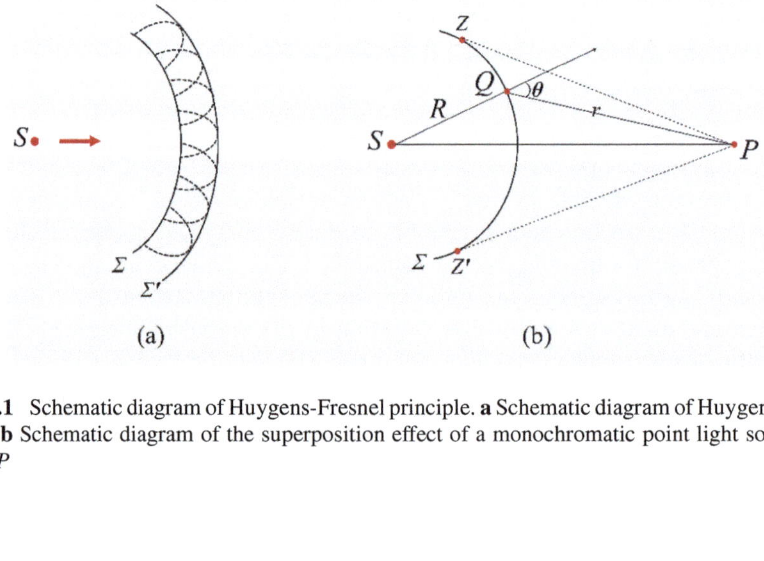

Fig. 2.1 Schematic diagram of Huygens-Fresnel principle. **a** Schematic diagram of Huygens' principle; **b** Schematic diagram of the superposition effect of a monochromatic point light source on point P

where $\tilde{\boldsymbol{E}}(\mathrm{P})$ represents the complex amplitude of the monochromatic point source S propagating to point P; $\mathrm{C} = 1/j\lambda$ is a constant, where j is the imaginary unit and λ is the wavelength of light in vacuum; $\tilde{\boldsymbol{E}}_{\mathrm{Q}}$ is the complex amplitude at point Q on the wavefront; r is the distance from point Q to point P; k_0 is the wave number of light propagating in vacuum; $K(\theta) = (1 + \cos\theta)/2$, where θ is the angle between the normal line at point Q on the wavefront and the line connecting QP $(-\pi/2 \leq \theta \leq \pi/2)$, as shown in Fig. 2.1b. By applying Eq. 2.1 across the entire wavefront Σ, we can derive the superposition outcome of all secondary waves emanating from the secondary sources located on the Σ surface at point P. This is encapsulated in Eq. 2.2, serving as the mathematical formulation of the Huygens-Fresnel principle.

$$\widetilde{\boldsymbol{E}}(P) = C \iint_{\Sigma} \widetilde{\boldsymbol{E}}_Q \frac{e^{ikr}}{r} K(\theta) d\sigma \tag{2.2}$$

where dσ represents the elementary wavefront at point Q.

As shown in Fig. 2.2, the dimensions of the silicon-based slab waveguide in the length (x-axis direction) and width (y-axis direction) are much larger than its height (z-axis direction). In this case, the region enclosed by the red dashed box in Fig. 2.2a is displayed in Fig. 2.2b, where the two yellow dashed lines at $x = x1$ and $x = x2$ represent the optical field distributions at the cross-sections of the silicon-based slab waveguide at $x = x1$ and $x = x2$, respectively. Assuming that the optical field at $x = x1$ is composed of N secondary wave sources, with the complex amplitude of any secondary wave source q being $\tilde{\boldsymbol{E}}_q$, the complex amplitude $\tilde{\boldsymbol{E}}_p$ of any secondary wave source p at $x = x2$ can be obtained using Eq. 2.3 inspired by the Huygens-Fresnel principle.

$$\widetilde{\boldsymbol{E}}_p = \frac{1}{j\lambda} \cdot \sum_{i=1}^{N} \left(\widetilde{\boldsymbol{E}}_{q=i} \cdot \frac{1 + \cos\theta_{pq}}{2r_{pq}} \cdot e^{jkr} \right) \tag{2.3}$$

where $k = (2\pi n_{\mathrm{slab}})/\lambda$, which is the wave number of light propagating in the silicon-based slab waveguide, r_{pq} is the distance between point p and point q, $\cos\theta$ represents the cosine of the angle between the line connecting points q and p and the line perpendicular to the cross-sections at $x = x_1$ and $x = x_2$, and n_{slab} is the effective refractive index of the silicon-based slab waveguide.

According to Eq. 2.3, the analytical expression for the light wave propagation between any two points at different positions in a silicon-based slab waveguide can be obtained, as shown in Eq. 2.4.

$$w_{pq}^{x=m} = \frac{1}{j\lambda} \cdot \left(\frac{1 + \cos\theta_{pq}}{2r_{pq}} \right) \cdot \exp\left(j \frac{2\pi r_{pq} n_{slab}}{\lambda} \right) \tag{2.4}$$

Here, $x = m$ represents the m-th cross-section of the optical field along the direction of light propagation; p represents a secondary wave source on the cross-section at $x = m$ and q represents a secondary wave source on the cross-section at $x = m-1$. Therefore,

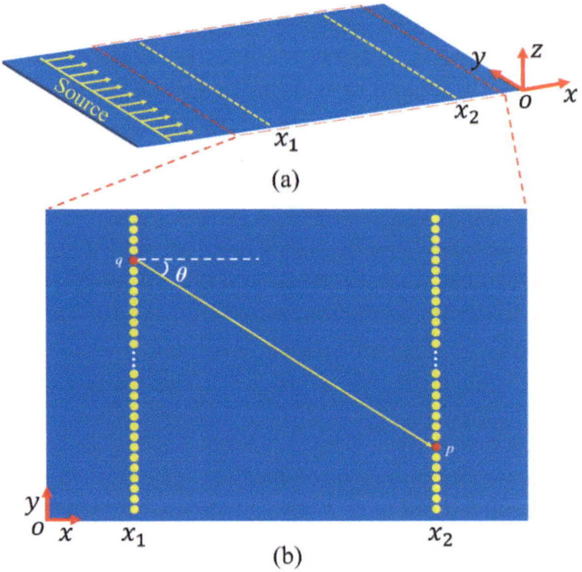

Fig. 2.2 Schematic diagram of silicon-based slab waveguide structure. **a** Schematic diagram of 3D silicon-based slab waveguide structure; **b** Schematic diagram of *xoy* plane of silicon-based slab waveguide

the distribution of any optical field in a silicon-based slab waveguide can be calculated according to Eq. 2.4. Subsequently, the 2.5D FDTD optical simulation software is used to verify the effect of Eq. 2.4 on the analytical expression of the optical field propagation when applied to the determined silicon-based slab waveguides.

To elucidate the propagation process of light in a silicon-based slab waveguide using Eq. 2.4, the effective refractive index n_{slab} of the waveguide is indispensable. Therefore, the effective refractive index of the silicon-based slab waveguide with different sizes was numerically calculated using optical simulation software (MODE Solutions). Figure 2.3 shows the calculated effective refractive index of the silicon-based slab waveguide with a thickness of 220 nm and various widths (0.45 μm ~ 105 μm). When the width of the silicon-based slab waveguide is greater than 10 μm, the calculated effective refractive index tends to stabilize around 2.8478.

Figure 2.4 illustrates the simulation conducted on a silicon-based slab waveguide with a width of 300 μm using 2.5D FDTD software. The outcomes of this simulation are displayed in Fig. 2.4a. From this simulation, the optical field characteristics at two cross-sections along the propagation direction, specifically at $x = 5$ μm and $x = 255$ μm, were meticulously extracted. The phase and amplitude distributions at these points are depicted in Fig. 2.4b through e. After extracting the phase and amplitude data at x = 5 μm, which are shown in Fig. 2.4b and d, these data were used as input parameters. Then, by utilizing Eq. 2.4, we computed the optical field distribution after a propagation distance of 250 μm. This computation aimed to replicate the optical field distribution at the $x = 255$ μm cross-section of the silicon-based slab waveguide, which was obtained from the 2.5D FDTD simulation. The results of this calculation are showcased in Fig. 2.5.

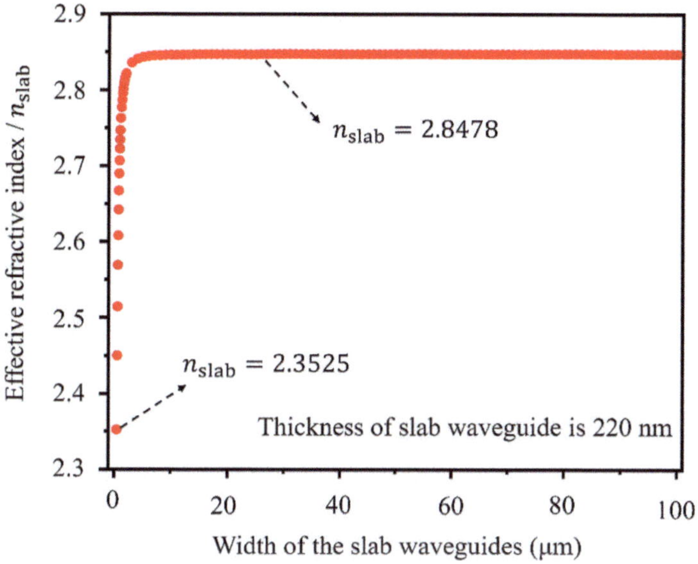

Fig. 2.3 Effective refractive index of silicon-based slab waveguides with different widths

In Fig. 2.5, the phase and amplitude distributions, as calculated using Eq. 2.4, exhibit discrepancies when compared to the results derived from the 2.5D FDTD software. Through iterative simulation and analysis, it was discerned that the outcomes from Eq. 2.4 diverge from those of the 2.5D FDTD software by an amplitude coefficient and a phase difference. Consequently, Eq. 2.4 was refined and augmented to yield Eq. 2.5, which is referred to as the two-dimensional silicon-based SDEPM, where A represents an amplitude coefficient and $\Delta\phi$ signifies a phase difference. Utilizing the phase and amplitude information at $x = 5$ μm once more as input signals, and integrating them with Eq. 2.5 for computation, a comparison was made between the calculated results and those obtained from the 2.5D FDTD optical simulation. This comparison, illustrated in Fig. 2.6, demonstrates a concordance between the results from both methodologies.

$$w_{pq}^{x=m} = \frac{1}{j\lambda} \cdot \left(\frac{1 + \cos\theta_{pq}}{2r_{pq}} \right) \cdot \exp\left(j\frac{2\pi\, r_{pq} n_{slab}}{\lambda} \right) \cdot A\exp(j\Delta\phi) \tag{2.5}$$

To further verify the effectiveness of Eq. 2.5 (i.e., the two-dimensional silicon-based SDEPM), additional silicon-based slab waveguides with different dimensions were selected for validation. The results are shown in Figs. 2.7 and 2.8. The calculations using the two-dimensional silicon-based SDEPM are consistent with those obtained from the 2.5D FDTD optical simulations. Thus, an analytical model (SDEPM) that can approximately characterize light propagation in silicon-based slab waveguides has been developed. The amplitude compensation factor A and the phase

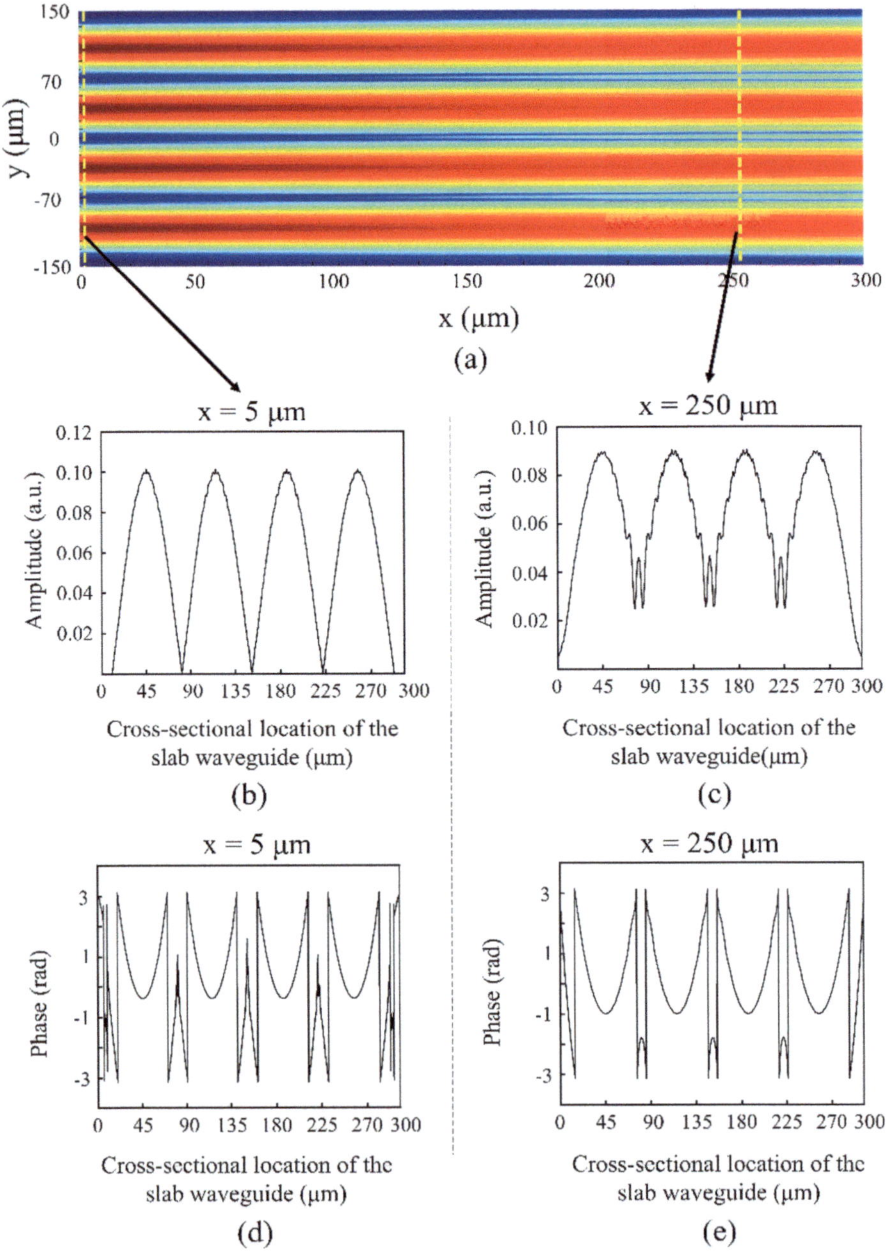

Fig. 2.4 Phase and amplitude distribution of the optical field in a silicon-based slab waveguide simulated by 2.5D FDTD. **a** 2.5D FDTD simulation results of the silicon-based slab waveguide simulation structure with a width of 300 μm and a length of 300 μm; **b** and **d** are the amplitude and phase distributions of the optical field at the cross-section of the silicon-based slab waveguide at $x = 5$ μm (the yellow dashed line in (a) represents the cross-section); **c** and **e** are the amplitude and phase distributions of the optical field at the cross-section of the silicon-based slab waveguide at $x = 255$ μm (the yellow dashed line in (a) represents the cross-section)

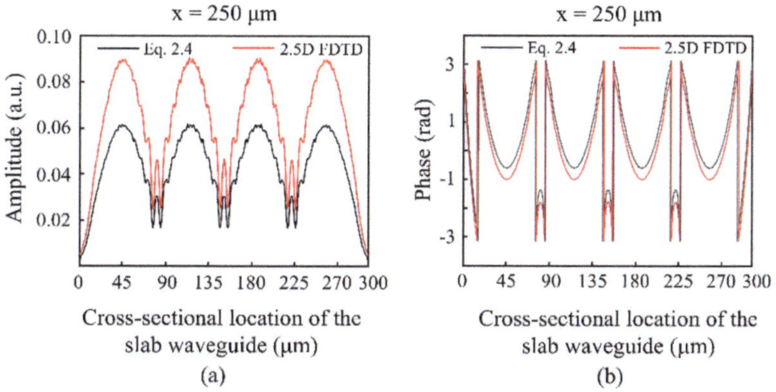

Fig. 2.5 Comparison of optical field propagation calculation results for the same input signal using Eqs. 2.4 and 2.5D FDTD software. **a/b** The black curve represents the phase/amplitude calculation result of the input signal propagating 250 μm in the slab waveguide (width 105 μm, length 300 μm, height 220 nm) using Eq. 2.4, while the red curve represents the phase/amplitude calculation result of the input signal propagating 250 μm in the slab waveguide using 2.5D FDTD software

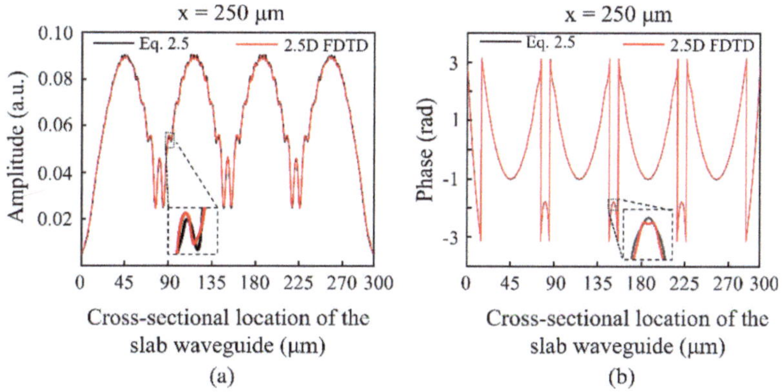

Fig. 2.6 Comparison of optical field propagation calculation results for the same input signal using Eqs. 2.5 and 2.5D FDTD software. **a** and **b**: The black curve represents the amplitude/phase calculation result of the input signal propagating 250 μm in the slab waveguide (width 105 μm, length 300 μm, height 220 nm) using Eq. 2.5; while the red curve represents the amplitude/phase calculation result of the input signal propagating 250 μm in the slab waveguide using 2.5D FDTD software

compensation factor $\Delta\phi$ in the two-dimensional silicon-based SDEPM can be determined by comparing the results from the 2.5D FDTD optical simulation software for specific silicon-based slab waveguide structures. The amplitude compensation factor A is introduced to compensate for errors arising from the discretization process. The phase compensation factor $\Delta\phi$ is included because the effective refractive index used in the two-dimensional silicon-based SDEPM is approximatively calculated by the optical simulation software MODE Solutions, which results in a difference compared to the effective refractive index inherently used in 2.5D FDTD software, leading to additional phase errors during calculation. For example, even if the effective refractive index used in the two-dimensional silicon-based SDEPM differs from that of 2.5D FDTD software by only 0.001 (i.e., on the order of 10^{-3}), a propagation distance of just 100 μm can result in a phase error of approximately 23° between the two calculation methods. Longer propagation distances would lead to even more significant phase differences. Therefore, the phase compensation factor $\Delta\phi$ is introduced in the two-dimensional silicon-based SDEPM to account for this discrepancy.

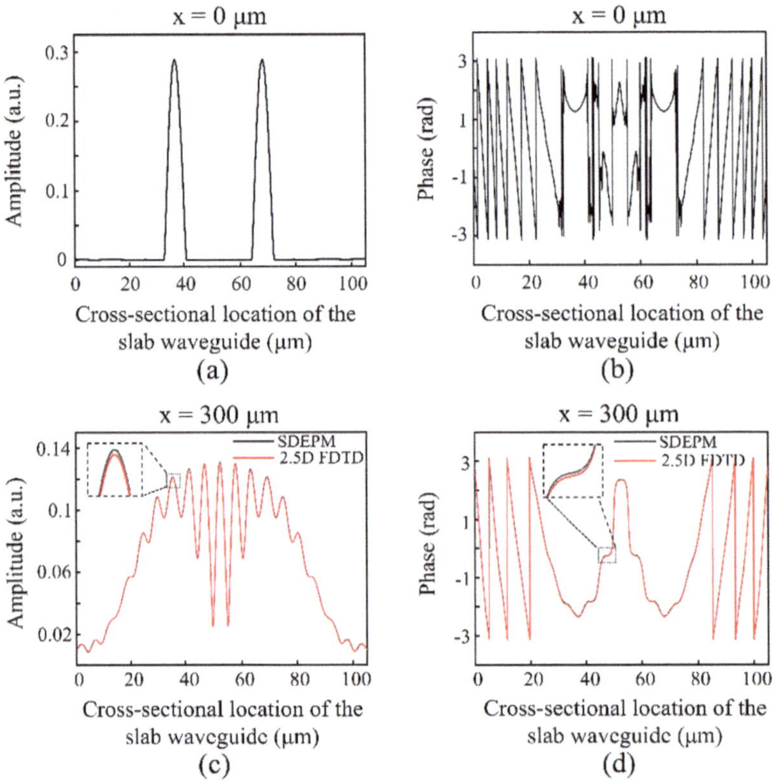

Fig. 2.7 Comparison of the optical field propagation calculation results for the same input signal between SDEPM and 2.5D FDTD model [1]. **a** and **b** are the amplitude and phase distributions of the input signal. **c** and **d** are the calculation results of the amplitude and phase of the input signal after it has propagated 300 μm in a two-dimensional silicon-based SDEPM (black curve) and 2.5D FDTD (red curve) in a slab waveguide (width 105 μm, length 300 μm, height 220 nm)

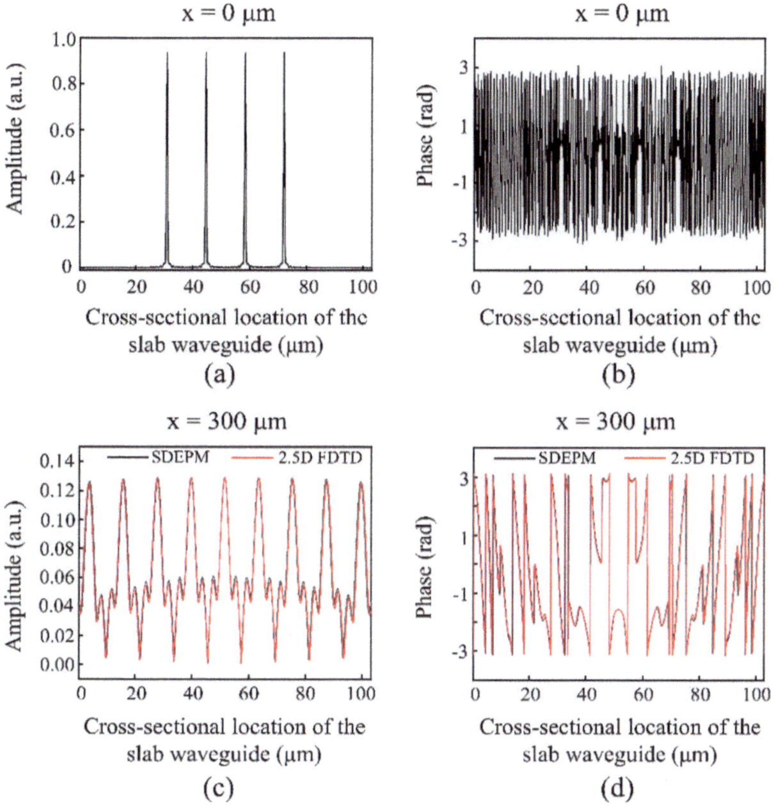

Fig. 2.8 Comparison of the calculation results of optical field propagation of SDEPM and 2.5D FDTD model for the same input signal [2]. **a** and **b** are the amplitude and phase distributions of the input signal; **c** and **d** are the calculation results of the amplitude and phase of the input signal after it has propagated 300 μm in two-dimensional silicon-based SDEPM (black curve) and 2.5D FDTD (red curve) in a slab waveguide (width 105 μm, length 300 μm, height 220 nm)

2.2 Diffractive Optical Neural Network Training Model

This section focuses on the training model for silicon-based integrated DONNs. Based on the two-dimensional silicon-based SDEPM proposed in Sect. 2.1, we can obtain an analytical expression for the forward propagation process while training silicon-based integrated DONNs. Therefore, algorithms such as stochastic gradient descent and adaptive moment estimation can be employed to pre-train the structure parameters of silicon-based integrated DONNs on a computer. According to the two-dimensional silicon-based SDEPM, namely Eq. 2.5, each neuron (subwavelength micro-slot unit) in the hidden layer of the silicon-based integrated DONN can be regarded as a secondary wave source. The amplitude and phase of this secondary wave source are determined by the input wave of the neuron and its own transmission coefficient (T). Therefore, the output function of the i-th neuron (x_i, y_i) in the m-th hidden layer of the silicon-based integrated DONN can be calculated using Eq. 2.6.

$$n_i^m(x_i, y_i) = w_i^m(x_i, y_i) \cdot T_i^m(x_i, y_i) \cdot \sum\nolimits_k n_k^{m-1}(x_i, y_i) \tag{2.6}$$

Here, $\sum\limits_k n_k^{m-1}(x_i, y_i)$ represents the input wave source of the i-th neuron at the position (x_i, y_i) in the m-th hidden layer of the silicon-based integrated DONN. After traveling a certain distance, the input wave source can reach the i-th neuron on the m-th hidden layer. This propagation process can be analytically expressed by Eq. 2.5, namely $w_i^m(x_i, y_i)$; $T_i^m(x_i, y_i)$ is the transmission coefficient of the i-th neuron on the m-th hidden layer, which can be expressed by Eq. 2.7.

$$T_i^m(x_i, y_i) = a_i^m(x_i, y_i) \cdot exp(j\varphi(x_i, y_i)) \tag{2.7}$$

where $a_i^m(x_i, y_i)$ denotes the amplitude coefficient, and as the neurons are represented by subwavelength micro-slot units with very small size, the transmission rate of the subwavelength micro-slot structure is very high. This implies that the energy loss of light traversing the subwavelength micro-slot array is minimal. Consequently, in this study, we set $a_i^m = 1$, meanwhile, $\varphi(x_i, y_i)$ represents the phase term of the i-th neuron on the m-th hidden layer of the silicon-based integrated DONN. Based on Eqs. 2.6 and 2.7, the forward propagation process in the training process of silicon-based integrated DONNs can be obtained, as shown in Eq. 2.8.

$$\begin{cases} n_{i,p}^m = w_{i,p}^m \cdot T_i^m \cdot u_i^m \\ u_i^m = \sum\nolimits_k n_{k,i}^{m-1} \\ T_i^m = a_i^m \cdot \exp(j\varphi_i^m) \end{cases} \tag{2.8}$$

Specifically, i denotes the i-th neuron on the m-th hidden layer of the silicon-based integrated DONN, and p denotes the p-th neuron on the next hidden layer, which is connected to the i-th neuron via diffraction.

Assuming the silicon-based integrated DONN comprises an M-layer network, excluding the input and output layers, the complex field and power of the p-th neuron on the output layer can be determined through Eqs. 2.9 and 2.10, respectively. Upon acquiring the output optical power, the discrepancy between the actual optical power $\left(S_i^{M+1}\right)$ and the ideal optical power $\left(Tar_k^{M+1}\right)$ can be quantified using the normalized mean square error method. This discrepancy is then iteratively minimized to approach as closely as possible to zero. The ideal optical power value, referred to here, serves as the target label. The mathematical expression facilitating this optimization endeavor is designated as the loss function (L), as delineated in Eq. 2.11.

$$u_i^{M+1} = \sum\nolimits_k n_{k,i}^M \tag{2.9}$$

$$S_i^{M+1} = \left|u_i^{M+1}\right|^2 \tag{2.10}$$

$$L(\varphi_i^m) = \frac{1}{N}\sum_k \left(\frac{S_k^{M+1}}{\sum_k S_k^{M+1}} - Tar_k^{M+1}\right)^2 \tag{2.11}$$

Specifically, N represents the total number of data points involved in the computation on the output interface. According to Eq. 2.11, the primary objective throughout the training regimen of the silicon-based integrated DONN's structural parameters is to minimize the absolute value of the loss function $L(\varphi_i^m)$, where φ_i^m signifies the phase term, bounded within the interval of $0 \sim 2\pi$. During the training process of the neural network structure parameters, the loss function value can be progressively reduced via gradient backpropagation. This investigation employs the chain rule to differentiate the phase term, as delineated in Eqs. 2.12 and 2.13.

$$\frac{\partial L}{\partial \varphi_i^m} = \frac{4}{N}\sum_k \left(\frac{S_k^{M+1}}{\sum_k S_k^{M+1}} - Tar_k^{M+1}\right) \cdot \frac{\sum_k S_k^{M+1} - S_k^{M+1}}{\left(\sum_k S_k^{M+1}\right)^2} \cdot \mathrm{Real}\left\{\left(u_k^{M+1}\right)^* \cdot \frac{\partial u_k^{M+1}}{\partial \varphi_i^m}\right\} \tag{2.12}$$

$$\frac{\partial u_k^{M+1}}{\partial \varphi_i^{m=M-L}} = j \cdot T_i^{M-L} \cdot u_i^{M-L} \cdot \sum_{k_1} w_{k_1,k}^M \cdot T_{k_1}^M \cdots\cdots \sum_{k_L} w_{k_L,k_{L-1}}^{M-L+1} \cdot T_{k_L}^{M-L+1} \cdot w_{i,k_L}^{M-L} \tag{2.13}$$

where m <M; $3 \leq L \leq M$.

In accordance with the network training model, each hidden layer's parameters (i.e., phase weight parameters) can be optimized by applying the forward propagation algorithm coupled with the error backpropagation algorithm. Before the initiation of the network structure parameters' training, it is imperative to procure an optical dataset. Therefore, the first step is to perform data preprocessing based on the characteristics of the target task dataset. In this study, data preprocessing encompasses two primary techniques: category-specific feature normalization and global feature normalization. Subsequently, the normalized data is scaled by a factor of π or 2π to map the normalized sample features onto the phase of light. It is noteworthy that beyond the phase of light, the sample features of the dataset can also be projected onto other optical properties, including intensity, polarization, angular momentum, and frequency. Next, the specific training process of the silicon-based integrated DONN structure parameters will be detailed, serving as the training procedure in the subsequent chapters.

This subsection will illustrate the methods of dataset preprocessing using the Iris dataset [3]. As shown in Fig. 2.9, the sepal length and width, as well as the petal length and width of the Iris flowers, are collected and used as the original Iris dataset. This dataset includes three species of Iris: Setosa, Versicolor, and Virginica, with 150 samples. Each sample has four features: sepal length, sepal width, petal length, and petal width.

During the data processing, the four feature values of each sample are first globally normalized. As shown in Fig. 2.9, after global normalization, the first sample features of Setosa (5.1, 3.5, 1.4, 0.2) are transformed into (0.6410, 0.4359, 0.1667, 0.0128). Subsequently, the globally normalized data are multiplied by 2π to map the sample features onto the phase of light. At this point, the phase feature values of the sample are obtained as (4.028, 2.739, 1.047, 0.08 rad). The original Iris dataset has been

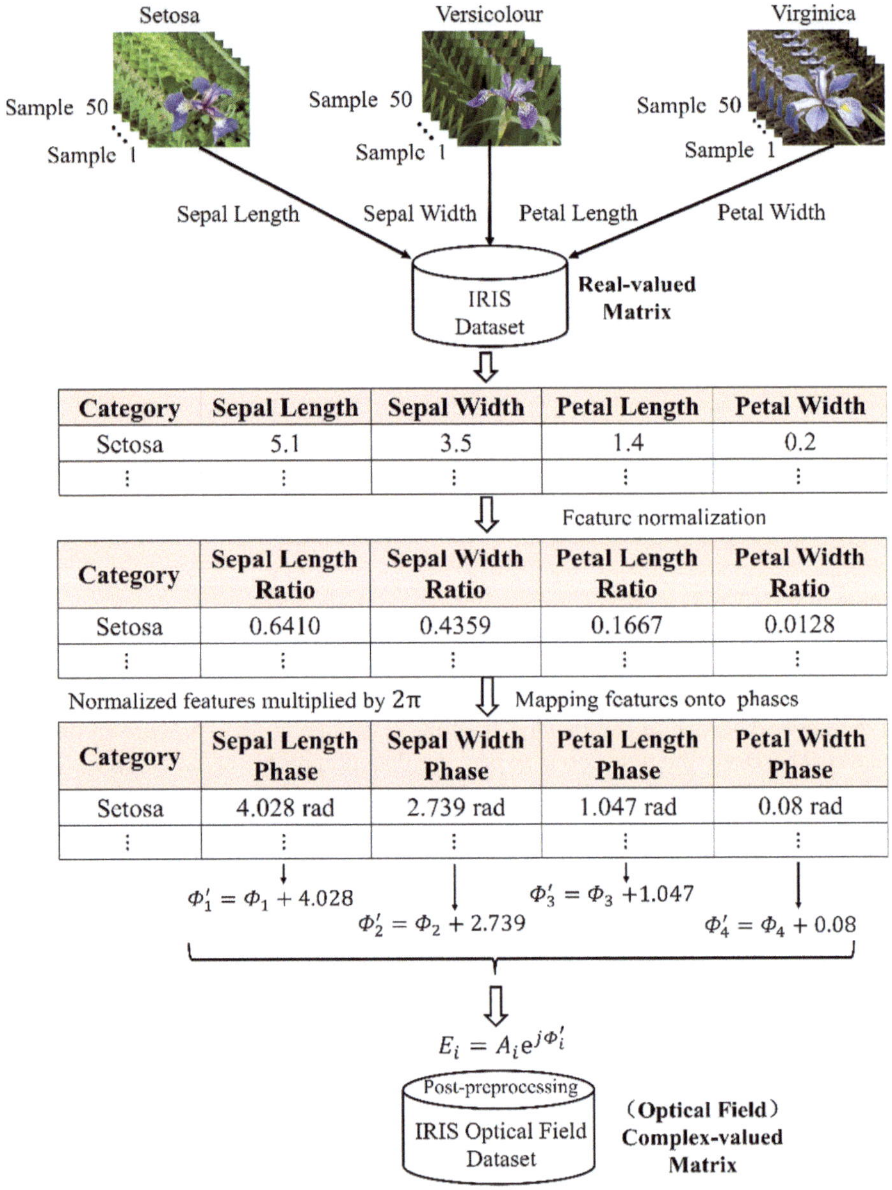

Fig. 2.9 Flowchart of dataset preprocessing

mapped onto the phase of light, and a phase-encoded Iris dataset with phase features has been obtained. Next, the Iris phase dataset will be loaded onto the initial phases of the silicon-based waveguides. In this example, the Iris dataset has four features, so there are four silicon-based input waveguides, with initial phases denoted as Φ_1, Φ_2, Φ_3 and Φ_4. The next step is to load the features of the Iris phase dataset onto the corresponding initial phases of the silicon-based input waveguides, such as Φ_1+4.028, Φ_2+2.739, Φ_3+1.047, Φ_4+0.08. After loading the features of the Iris phase dataset onto the initial phases of the silicon-based input waveguides, new phase data are obtained in the form of ($\Phi_1\prime$, $\Phi_2\prime$, $\Phi_3\prime$, $\Phi_4\prime$). These phase data are then combined with the corresponding amplitude fields in the silicon-based input waveguides (e.g., $E_i = A_i e^{j\Phi_i\prime}$) to obtain the preprocessed Iris optical field dataset (a complex-valued matrix). At this time, each sample in the dataset represents an optical field distribution, which serves as the training dataset for the key parameters of the silicon-based integrated DONNs. The phase and amplitude fields of the silicon-based input waveguides can be obtained by using 2.5D FDTD software simulation. The optical field of the input waveguide needs to be extracted only once before the network training begins and does not need to be repeatedly acquired.

After obtaining the optical field dataset, the training set from this dataset is fed into the training network for optimizing the structure parameters of silicon-based integrated DONNs, as shown in Fig. 2.10. Initially, the phase weight parameters on the first hidden layer (HL1) are randomly initialized. When the input optical field passes through the two-dimensional silicon-based SDEPM and reaches HL1, the optical field is summed at the front of each complex-valued neuron on HL1. Subsequently, the summed optical field interacts with the corresponding complex-valued neuron on HL1. After this operation is completed, the output field of the entire HL1 serves as the input field (secondary wave source) for the second hidden layer (HL2), where similar operations are performed. This process continues sequentially. Each hidden layer's output field serves as the input wave for the next hidden layer. This way, the output field from the last hidden layer (HLm, where m represents the m-th hidden layer in the silicon-based integrated DONN) passes through the two-dimensional silicon-based SDEPM. It reaches a specific region at the output interface of the silicon-based integrated DONN, where power summation of the optical field is performed. By calculating the normalized mean squared error (NMSE), the summed optical power is then normalized and compared with the corresponding label for the input sample. The NMSE values for all samples are summed and averaged. The averaged NMSE value is compared with the target set value. If the requirement is not met, the phase weight parameters on the hidden layers are updated layer-by-layer, and the training process described above is repeated until the average NMSE for all samples reaches the target set value. Finally, the optimal structure parameters for this silicon-based integrated DONN training session are obtained, and the training is concluded.

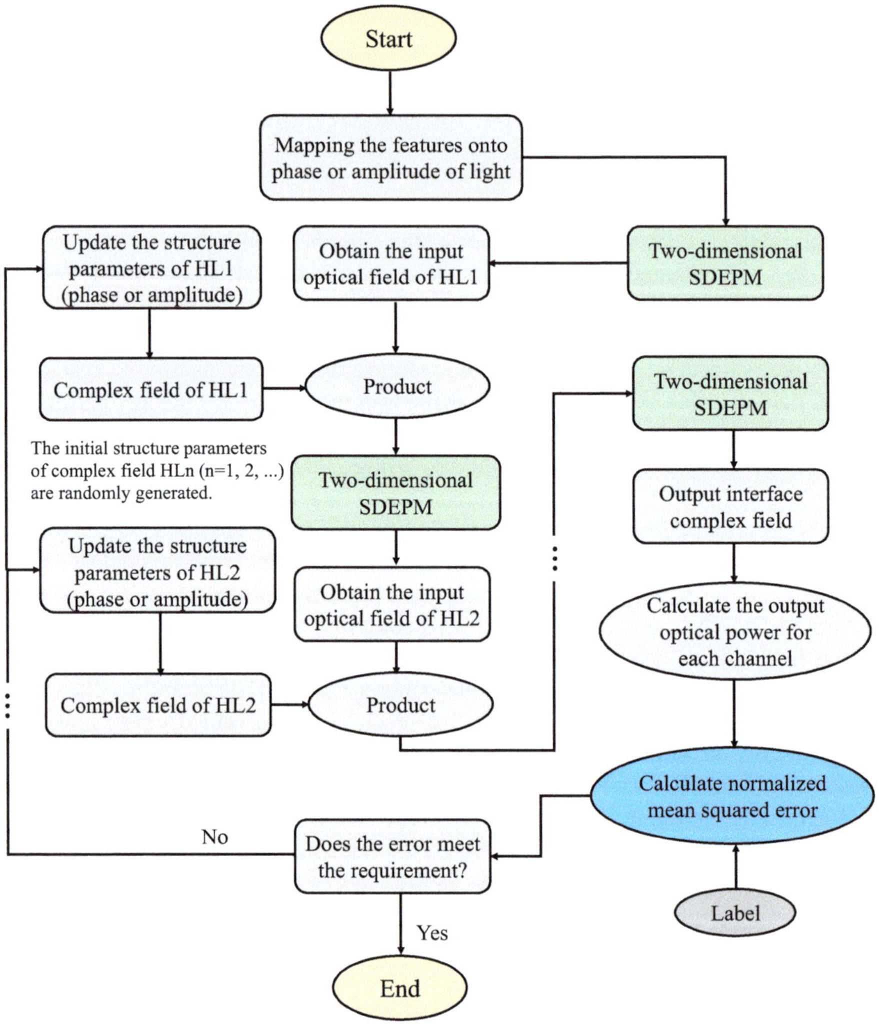

Fig. 2.10 Flowchart of training the structure parameters of silicon-based integrated DONN. Here, HLm represents the m-th hidden layer of the DONN. For example, HL1 and HL2 represent the first and second hidden layers of the DONN, respectively

2.3 Phase Weight Mapping Model

The optimizing process and methods for the parameters of the silicon-based integrated DONN have already been discussed in Sect. 2.2. It is crucial to emphasize that, before training the neural network, it must be ensured that all structure parameters used in the pre-training stage on the computer have practical physical significance. Otherwise, the silicon-based integrated DONN chip fabricated based on the optimal parameters obtained from the pre-training stage will not achieve the corresponding neural network inference functions. In this research, the primary physical quantity involved in the optimization and training of the silicon-based integrated DONN structure parameters is the phase of light. It has been analyzed that the physical medium of SDEPM throughout the training process is a silicon-based slab waveguide with a width greater than 10 μm. According to Fig. 2.3, the effective refractive index of the slab waveguide is 2.8478. However, when employing a silicon slot structure filled with silicon dioxide (SiO_2) material (SSSD) for phase weight mapping implementation, it is essential to establish and characterize the precise relationship between the induced phase shift and the geometric dimensions of the SSSD. Otherwise, it is impossible to update or optimize the structure parameters of the silicon-based integrated DONN on a computer. Equation 2.14 is employed to determine the length of the SSSD using the phase weight parameter.

$$L_{\text{slot}-i} = \frac{\Delta\varphi_i}{(n_{\text{eff}} - n_{\text{slab}}) \cdot k_0} \tag{2.14}$$

where $\Delta\varphi_i$ is the phase weight parameter (phase shift) obtained through optimization; $L_{\text{slot}-i}$ is the length of the SSSD determined under the i-th phase parameter; n_{eff} is the effective refractive index of the SSSD region; n_{slab} is the effective refractive index of the silicon-based slab waveguide; and k_0 represents the wave number of light propagating in vacuum.

In Eq. 2.14, there is a critical correlation parameter between $\Delta\varphi_i$ and $L_{\text{slot}-i}$, which is the effective refractive index n_{eff} of the SSSD region. Therefore, the specific value of n_{eff} needs to be determined before training the structure parameters of the silicon-based integrated DONN. In our approach, we first calculate the effective refractive index of the SSSD array by applying the Effective Medium Theory (EMT) [4, 5]. The EMT treats the periodic SSSD array as a uniaxial crystal. In this study, the center-to-center spacing between adjacent SSSD on thc silicon slab waveguide is 500 nm. The SSSD array can be regarded as a diffraction grating with sub-wavelength structure dimensions, thus satisfying the conditions for calculating the effective refractive index of a uniaxial crystal using the EMT approximation.

When the direction of the electric field is perpendicular to the grating vector of the SSSD array, the structure's zero-order and second-order approximations of the effective permittivity are shown in Eqs. 2.15 and 2.16, respectively.

$$\varepsilon_{\perp}^{(0)} = f\varepsilon_{\text{SiO}_2} + (1-f)f\varepsilon_{\text{Si}\perp} \tag{2.15}$$

$$\varepsilon_{\perp}^{(2)} = \varepsilon_{\perp}^{(0)}\left[1 + \frac{\pi^2}{3}\left(\frac{\Lambda}{\lambda}\right)^2 f^2(1-f)^2 \frac{\left(\varepsilon_{\text{Si}\perp} - \varepsilon_{\text{SiO}_2\perp}\right)^2}{\varepsilon_0\varepsilon_{\perp}^{(0)}}\right] \tag{2.16}$$

When the direction of the electric field is parallel to the grating vector of the SSSD array, the structure's zero-order and second-order approximations of the effective permittivity are shown in Eqs. 2.17 and 2.18, respectively.

$$\varepsilon_{\parallel}^{(0)} = \frac{\varepsilon_{SiO_2}\varepsilon_{Si\parallel}}{f\varepsilon_{Si\parallel} + (1-f)\varepsilon_{SiO_2}} \tag{2.17}$$

$$\varepsilon_{\parallel}^{(2)} = \varepsilon_{\parallel}^{(0)}\left[1 + \frac{\pi^2}{3}\left(\frac{\Lambda}{\lambda}\right)^2 f^2(1-f)^2\left(\varepsilon_{Si\parallel} - \varepsilon_{SiO_2}\right)^2 \frac{\varepsilon_{\perp}^{(0)}}{\varepsilon_0}\left(\frac{\varepsilon_{\parallel}^{(0)}}{\varepsilon_{Si\parallel}\varepsilon_{SiO_2}}\right)^2\right] \tag{2.18}$$

where λ is the wavelength of light in vacuum; f is the filling factor of the rectangular structure (in this study, the filling material of the rectangular micro-slot structure is SiO_2 material), $f = w_{slot}/d$, where w_{slot} is the width of the SSSD, and d is the lattice constant (i.e. the grating period Λ, the center-to-center distance between two adjacent SSSD structures) of the periodic SSSD array; $\varepsilon_{Si\perp}$ is the effective permittivity of the TM mode in the silicon-based slab waveguide; $\varepsilon_{Si\parallel}$ is the effective permittivity of the TE mode in the silicon-based slab waveguide; ε_{SiO_2} is the relative permittivity of the SiO_2 material in the rectangular micro-slot structure; $\varepsilon_{\perp}^{(0)}$ and $\varepsilon_{\parallel}^{(0)}$ represent the zero-order approximation of the effective permittivity; $\varepsilon_{\perp}^{(2)}$ and $\varepsilon_{\parallel}^{(2)}$ represent the second-order approximation of the effective permittivity. Since the direction of the electric field in the silicon-based slab waveguide is parallel to the grating vector, according to Eq. 2.18, the effective refractive index of SSSD regions with different widths when the period is 500 nm is obtained as shown in Fig. 2.11.

Secondly, by using the same structure parameters as those in the EMT calculation of the effective refractive index of the SSSD region (namely, the same SSSD length and period), the 2.5D FDTD software is used to calculate the effective refractive index of the SSSD region with different widths. The calculation method is to measure the phases at both ends of the SSSD and calculate their difference. For example, the phase values φ_{21} and φ_{22} at both ends of the slot are measured by the line monitor in Mode Solutions, and the phase shift $\Delta\varphi_2$ is calculated. Then, the effective refractive index of the corresponding SSSD region is obtained by using Eq. 2.14 (at this time, let

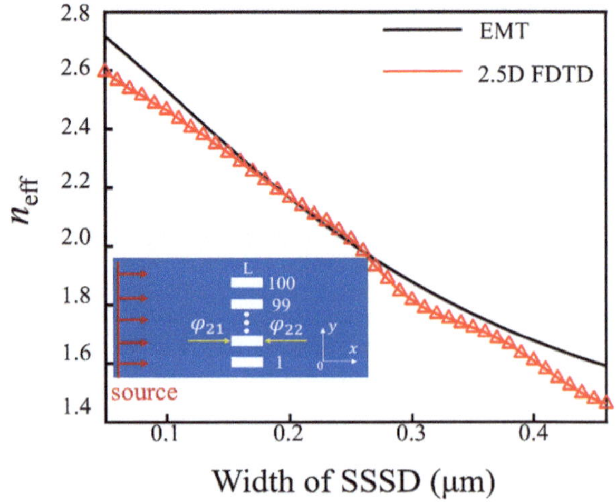

Fig. 2.11 Effective refractive index of SSSD regions with different widths. The black curve is the calculation results from EMT, and the curve marked with red triangles is from 2.5D FDTD software. The center-to-center distance between adjacent SSSDs is 500 nm

$n_{slab} = 0$). The calculation results are shown in Fig. 2.11. The black curve represents the calculation results of the effective refractive index of the SSSD region with different widths using EMT, and the curve marked with red triangles represents the calculation results of the effective refractive index of the SSSD region with different widths using 2.5D FDTD software.

Combining the calculation results of the effective refractive index of the SSSD region from the two methods of EMT and 2.5D FDTD and considering the fabrication conditions of the actual silicon-based integrated DONN chip, this study selects an SSSD with a width of 200 nm as the basic structure. The effective refractive index of the SSSD region at this width is 2.166. As shown in Fig. 2.12, the SSSD has a width of 200 nm and a length of 2 μm. Using 2.5D FDTD, the effective refractive index calculations for the SSSD regions with different distances from the light source (with spacings of 30 μm and 300 μm, respectively) all yield a result of 2.166. This result indicates that the effective refractive index of the SSSD region is stable when the plane light is incident.

When the width of the SSSD is fixed at 200 nm, the length of the SSSD within the SSSD array is varied, and the effective refractive index of the SSSD region, as well as the transmittance and phase shift of light propagating through the SSSD array, are calculated using 2.5D FDTD software. The calculation results are shown in Fig. 2.13. The transmittance of the SSSD array is greater than 94% and does not change significantly with the variation in SSSD length. Regarding the phase shift, a phase difference ranging from 0 to 2π can be achieved when the SSSD length varies from 0 to 2.3 μm. It is worth noting that, considering the precision limitations in the fabrication of physical devices, the phase shift achieved within the length range of 0 to 0.2 μm can be equivalently realized in a length range of 2.3 to 2.5 μm according to the 2 π phase wrapping theorem. Therefore, the length of the SSSD can be flexibly adjusted according to the specific conditions of the fabrication process.

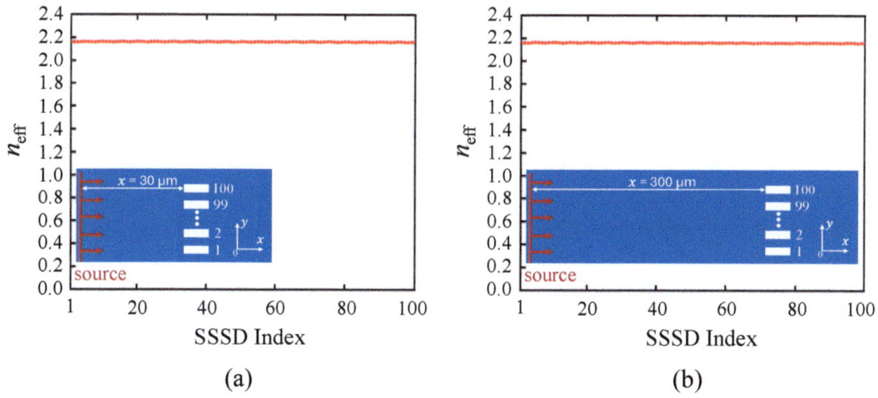

Fig. 2.12 Calculation results of the effective refractive index of the SSSD region at different distances (30–300 μm) between the light source and the SSSD array

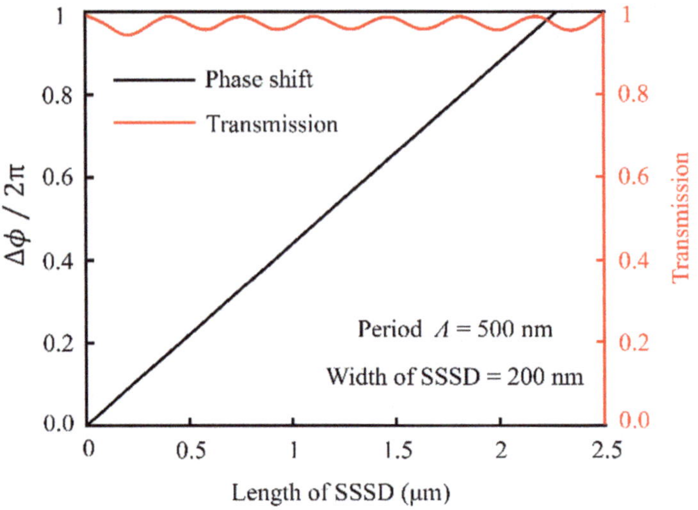

Fig. 2.13 Variation of the transmittance and phase shift of the SSSD array with various lengths

At this stage, we have theoretically established a precise quantitative correspondence between the phase weight parameters and the geometric dimensions of the SSSD, as mathematically formulated in Eq. 2.14. Under this condition, the parameter optimization process for the silicon-based integrated DONN can be carried out on the computer. Once the silicon-based integrated DONN structure parameters are obtained, the next step is to fabricate the DONN chip. Another important issue that needs to be considered is how to ensure that the phase weight parameters obtained through the pre-training stage on the computer can be accurately realized during the physical structure mapping process of the DONN chip. This is because the above discussion is based on the condition of plane light incident on a uniaxial crystal, where each SSSD in the SSSD array has the same length, width, and height. However, practically, after the training of the silicon-based integrated DONN structure parameters is completed, the obtained phase weights $\Delta\varphi_i$ are different from each other. The lengths of the SSSDs calculated by employing Eq. 2.14 will also be different. Therefore, it is necessary to verify further whether the correspondence between the phase weight parameters and the dimensions of the SSSD still holds. Moreover, in practical situations, the propagating light does not necessarily propagate horizontally to each SSSD. Furthermore, when the light wave passes through the first layer of SSSDs, diffraction occurs. The secondary wave sources generated continue to propagate to the next layer of the SSSD array, and the input angles at which these secondary wave sources propagate to each SSSD are also different. In this case, it is also necessary to verify whether the correspondence between the phase weight parameters and the dimensions of the SSSD still holds.

Figure 2.14a illustrates the simulation structure of 2.5D FDTD, where plane light is incident, and the lengths of each SSSD in the SSSD array are assigned randomly. Under this condition, the effective refractive index of 20 randomly selected SSSD

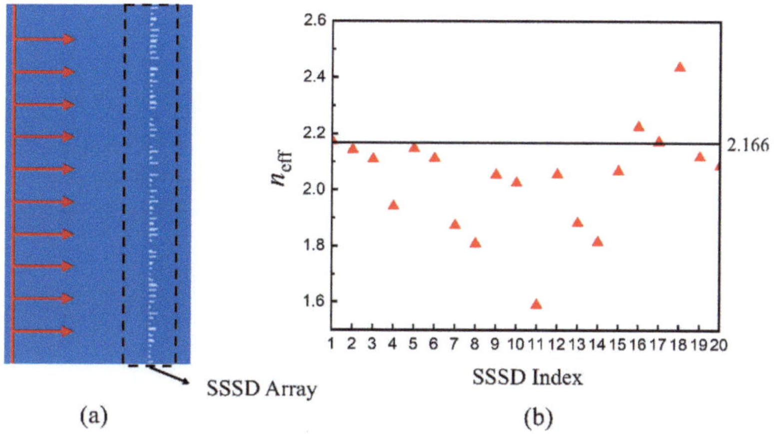

Fig. 2.14 Calculation results of the effective refractive index for 20 randomly selected SSSD regions in an SSSD array. **a** Plane light is incident on the SSSD array. As shown in the black dashed box, the SSSD array is composed of SSSDs with a period of 500 nm, a width of 200 nm, and different lengths; **b** The effective refractive index of 20 randomly selected SSSDs is presented in the black dashed box of Fig. 2.14a. The selected SSSDs are numbered from 1 to 20 from bottom to top

regions within the SSSD array, calculated using 2.5D FDTD software, is shown in Fig. 2.14b. The effective refractive index obtained here differs from the effective refractive index ($n_{\mathrm{eff0}}= 2.166$) used in the pre-training of phase weight parameters for the silicon-based integrated DONN on the computer. Therefore, the phase weight parameters obtained through the pre-training stage on the computer do not correspond to the phase shift realized by the SSSDs during the chip fabrication process. In other words, the silicon-based integrated DONN chip fabricated under these conditions will not possess the inference capabilities of the silicon-based integrated DONN obtained through the pre-training stage.

In Fig. 2.14b, the effective refractive index calculated using 2.5D FDTD software varies due to the different lengths of the SSSDs. Analysis reveals that this result is caused by the non-uniform dielectric distribution in the SSSD regions due to the varying lengths of adjacent SSSDs. To address this issue, multiple identical SSSDs of the same length are grouped (denoted as SSSDG) for phase weight mapping and the same method is used to calculate the effective refractive index of the SSSDG region in 2.5D FDTD. As shown in Fig. 2.15, when the number of SSSDs with the same dimensions in the SSSDG increases, the calculated effective refractive index of the SSSDG region tends to stabilize. It approaches the effective refractive index of 2.166 used during the pre-training stage on the computer. Only when this fundamental condition is satisfied can the phase shifts realized in the SSSDG regions after chip fabrication be consistent with the phase weight parameters obtained through pre-training stage on the computer. Therefore, using SSSDG to implement phase weight mapping can resolve the issue of unstable effective refractive index caused by non-uniform dielectric distribution between SSSDs of different lengths. Thereby ensuring

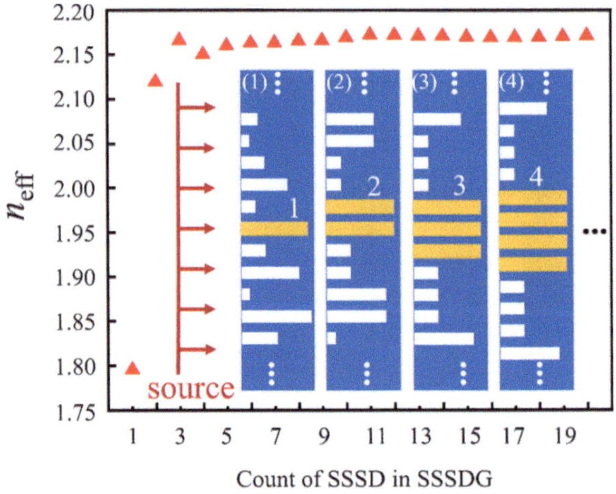

Fig. 2.15 Calculation results of the effective refractive index of SSSDG regions with various numbers of SSSDs. The yellow-marked SSSDs are the same size and form different SSSDGs respectively. When plane light is incident on the SSSD array, the effective refractive index of the SSSDG regions with different numbers of SSSDs is shown in the figure. The effective refractive index of 20 SSSDG regions, numbered 1–20, was calculated

that the fabricated silicon-based integrated DONN chip can achieve the inference capabilities of the DONN obtained through the pre-training stage on the computer.

The next issue that needs to be discussed is whether the effective refractive index of the SSSD region, calculated under non-plane light incidence conditions, remains close to the effective refractive index of 2.166 used during the pre-training stage on the computer. In this case, a single-mode waveguide with a TE_0 mode source is used as the incident light, and the effective refractive index of each SSSD region is calculated at different distances between the light source and the SSSD array. The calculation results are shown in Fig. 2.16. Analysis reveals that the smaller the incident angle of the light propagating to the SSSD, the closer the calculated effective refractive index of the SSSD region is to 2.166. Conversely, when the incident angle is larger, the calculated effective refractive index varies significantly.

To analyze more intuitively the impact of the incident angle on the effective refractive index calculation of the SSSD region, the effective refractive index calculation results of different numbered SSSDs (numbered from 1 to 100 in the y-direction from bottom to top in Fig. 2.16) in the SSSD array at the cross-sections of $x = 30$ μm, $x = 250$ μm, and $x = 450$ μm in Fig. 2.16 are displayed in Fig. 2.17. It can be observed that as the distance between the incident light source and the SSSD array increases, i.e., as the incident angle of the light propagating to each SSSD decreases, the effective refractive index of the SSSD region approaches 2.166 more closely. Therefore, when designing silicon-based integrated DONNs, it is advisable to consider increasing the distance between the incident light source and the hidden

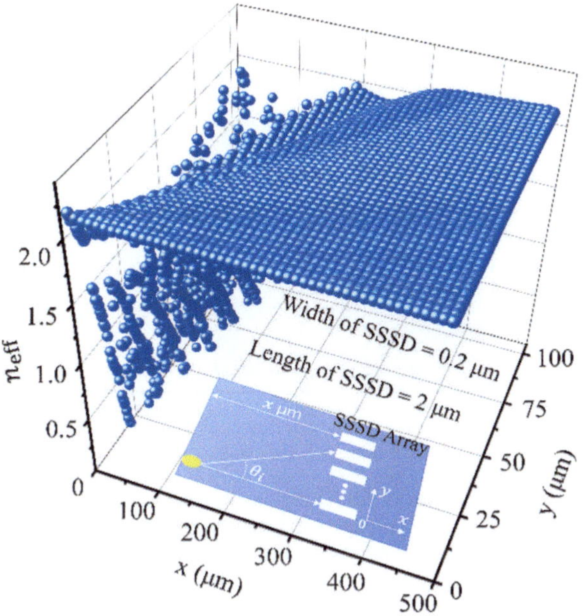

Fig. 2.16 Calculation results of the effective refractive index of SSSD regions at different distances between the light source and the SSSD array [1]. The yellow dot represents the position of the light source, and the positive x-axis represents the direction of light propagation. The length and width of each SSSD in the SSSD array are the same, with values of 2 μm and 0.2 μm, respectively. The SSSD array is arranged along the positive y-axis with a period of 0.5 μm. The blue sphere represents the calculated effective refractive index of each SSSD region at different distances between the light source and the SSSD array

layers (as well as between hidden layers) to achieve a more accurate mapping of phase weight parameters onto the chip's physical structure.

When designing silicon-based integrated DONNs, using 2.5D FDTD method, increasing the spacing between the incident light source and the hidden layer results in more stable effective refractive index calculations for the SSSD region. These calculations are also closer to the effective refractive index used in pre-training (n_{eff0}= 2.166). This approach enables more accurate mapping of the pre-trained phase weight parameters onto the physical structure of the DONN chip. As previously described, when calculating the effective refractive index of the SSSD region for various lengths using 2.5D FDTD, the method involves measuring the phase shift between the two ends of the SSSD and then solving for the effective refractive index of the corresponding SSSD region using Eq. 2.14. To illustrate why the effective refractive index calculation becomes more stable as the spacing between the input light source and the SSSD array increases, Fig. 2.18 is referred to here. The figure shows the wavefront distribution during the propagation of a point light source. In Fig. 2.18, identical SSSD arrays are placed at different propagation distances from the point light source, and each SSSD within the array has identical dimensions.

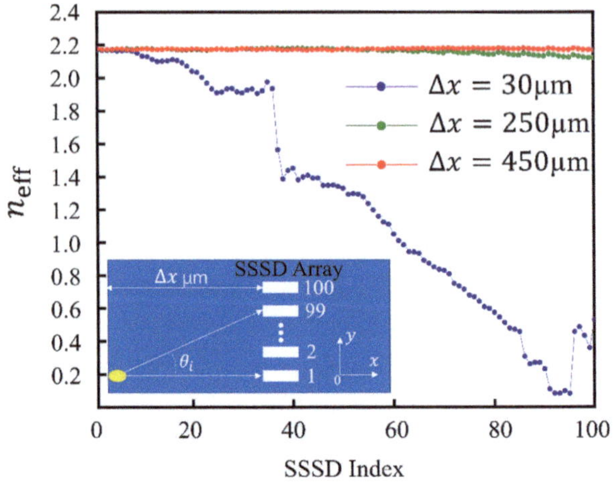

Fig. 2.17 Effective refractive index of SSSD regions in the SSSD array, calculated by the 2.5D FDTD method, for incident light at propagation distances of 30, 250, and 450 μm from the light source. The yellow elliptical mark represents the position of the light source, and the positive x-axis represents the direction of light propagation. The length and width of each SSSD in the SSSD array are 2 μm and 0.2 μm, respectively. The SSSDs are arranged along the positive y-axis with a period of 0.5 μm

The phase shift between adjacent wavefronts is constant, and the phase distribution on the same wavefront is uniform. Therefore, it can be inferred that, for the SSSD array closer to the light source, the phase shifts between the two ends of SSSDs are different. For example, the phase differences $\Delta\varphi_1$ and $\Delta\varphi_n$ are significantly different. Consequently, measuring the phase shifts between the two ends of each SSSD and solving for the effective refractive index of the corresponding SSSD region using Eq. 2.14 will yield different results. In contrast, for the SSSD array farther from the point light source, the wavefront distribution is relatively smooth due to the longer propagation distance of light. As shown in Fig. 2.18, the phase differences between the two ends of each SSSD are almost identical. For example, the phase differences $\Delta\varphi_1$ and $\Delta\varphi_\text{n}$ are very close this time. Therefore, measuring the phase difference between the two ends of each SSSD and solving for the effective refractive index of the corresponding SSSD region using Eq. 2.14 will yield nearly identical results, with small fluctuations around 2.166.

Based on the above discussion, we implemented the following phases, namely $(-\pi/4)$, $(-3\pi/4)$, and $(-5\pi/4)$, using SSSDGs with various lengths to achieve the corresponding phase shifts in the physical structures. Each SSSDG consists of 10 SSSDs with identical dimensions, and the length of each SSSD group is calculated based on Eq. 2.14. At this point, we designed a set of comparative simulation experiments. Both groups used the point light source (single-mode silicon waveguide with dimensions of 450 nm in width and 220 nm in height, TE_0 mode). Figure 2.19d, e, and f represent the experimental group, where the fundamental mode (TE_0) light source is 300 μm away from the SSSDG array. Figure 2.19a, b, and c represent the

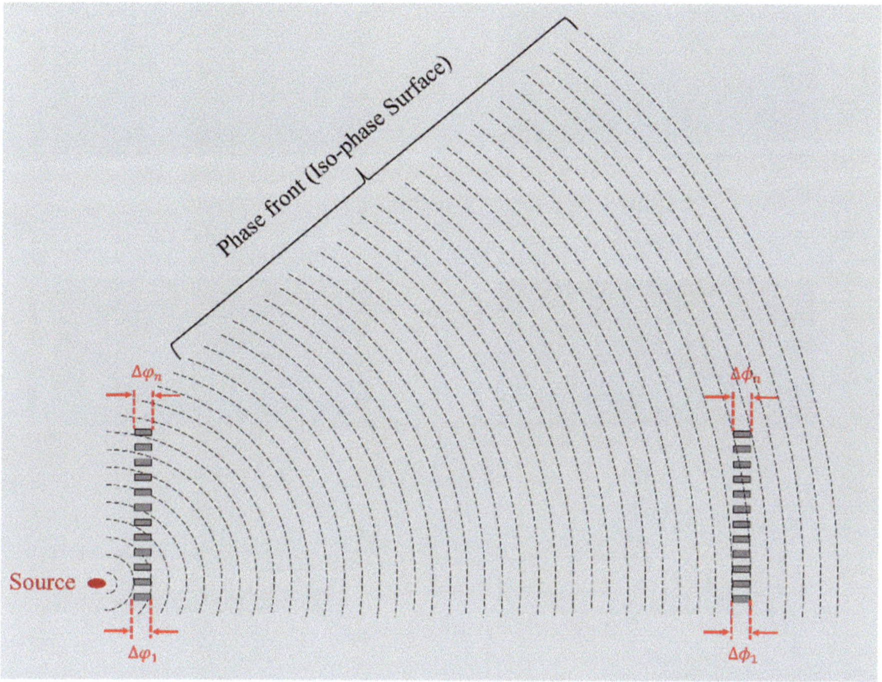

Fig. 2.18 Distribution of the wavefront of a point light source during the propagation process

comparison group, where the fundamental mode light source is 30 μm away from the SSSDG array.

As shown in Fig. 2.19c, when the light source is only 30 μm away from the SSSDG array, the phase shifts measured between the two ends of the SSSDG using 2.5D FDTD are completely different from the preset phase difference values of $(-\pi/4)$, $(-3\pi/4)$, and $(-5\pi/4)$. However, when the source is 300 μm away from the SSSDG array, the phase shifts measured between the two ends of the SSSDG using 2.5D FDTD are essentially the same as the preset phase difference values of $(-\pi/4)$, $(-3\pi/4)$, and $(-5\pi/4)$, as shown in Fig. 2.19f. By comparing the results of the experimental and comparison groups, it can be concluded that increasing the distance between the source and the SSSDG array ensures the accurate realization of the preset phase difference values in the physical structure. This method also applies to the spacing between hidden layers, as each SSSDG on a hidden layer can be regarded as a secondary wave source. When the distance between hidden layers is sufficiently large, the SSSDGs on the next hidden layer can accurately realize the preset phase difference values in the physical structure. As depicted in Fig. 2.19, the phase weight parameters derived from the pre-training stage can be precisely mapped onto the physical structure of the DONN chip when the following two conditions are satisfied concurrently: First, the SSSDG physical structure is employed to implement the phase weight parameters obtained in the pre-training

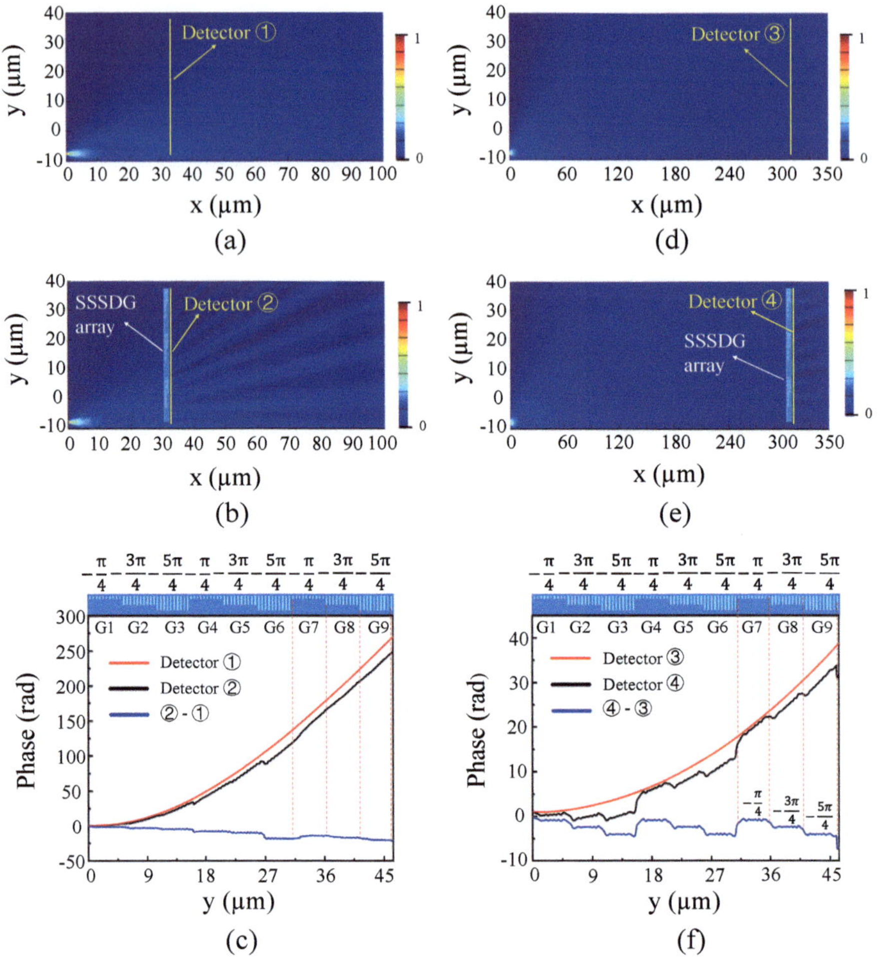

Fig. 2.19 Simulation verification results of SSSDG for predetermined phase values [1]. **a** and **d** show that the distances from the point source to monitors ① and ③ are 32 μm and 302 μm, respectively; **b** and **e** show that the same SSSDG array is located at distances of 30 μm and 300 μm from the point source, respectively, and that monitors ② and ④ are arranged 2 μm behind the SSSDG array; **c** The numerical analysis of the monitors in **a** and **b**; **f** The numerical analysis of the monitors in **d** and **e**. G1 to G9 are the numbers of different SSSDGs, and each SSSDG consists of 10 identical SSSDs

stage on the computer. Second, a certain distance is maintained between the incident light source and the SSSDG array to ensure that the incident light propagates to the SSSDG array at as small an angle as possible.

2.4 Summary

Section 2.1 begins with the Huygens-Fresnel principle to introduce the propagation process of light. Inspired by the Huygens-Fresnel diffraction equation, the propagation formula for optical field evolution in a silicon slab waveguide was derived, namely Eq. 2.5, which is referred to as the two-dimensional silicon-based SDEPM. Since the two-dimensional silicon-based SDEPM is an analytical expression, the forward propagation algorithm, gradient descent algorithm, and error backpropagation algorithm used in the neural network optimization can be directly applied to train the structure parameters of the silicon-based integrated DONNs on a computer. The training procedure for the DONNs is detailed in Sect. 2.2, and the training procedures for the structure parameters of the silicon-based integrated DONN in subsequent chapters will follow the same process as described in Sect. 2.2.

After obtaining the structure parameters of the silicon-based integrated DONN, another key issue to address is the mapping of phase weight parameters, that is, how to ensure that the DONN parameters obtained through pre-training on the computer are accurately realized in the physical structure of the DONN chip. Otherwise, the fabricated DONN chip will not possess the inference capabilities of the originally designed silicon-based integrated DONN. To make the mapping of phase weight parameters onto the physical structure of the chip more accurate, in Sect. 2.3, specific and effective measures are proposed. These measures include using SSSDG as the basic unit to implement the phase weight parameters obtained through the pre-training stage, and increasing the spacing between the incident light source and the SSSDG array so that the incident light propagates to each SSSDG region at a smaller incident angle. Finally, the effectiveness of the phase weight mapping model was demonstrated through simulation-based comparative experiments in the physical structure via 2.5D FDTD method.

At this point, the silicon-based integrated DONN model has been established, and all subsequent work in the following chapters will be based on this model. This chapter's content will play an important theoretical guiding role in the design and fabrication of silicon-based integrated DONN chips.

References

1. Fu T, Zang Y, Huang H et al (2021) On-chip photonic diffractive optical neural network based on a spatial domain electromagnetic propagation model. Opt Express 29(20):31924–31940
2. Fu T, Huang Y, Sun R et al (2023) Integrated diffractive optical neural network with space-time interleaving. Chin Opt Lett 21(9):091301
3. Blake CL (1998) UCI repository of machine learning databases. http://www.ics.uci.edu/~mlearn/MLRepository.html
4. Rytov S (1956) Electromagnetic properties of a finely stratified medium. Soviet Phys JEPT 2:466–475
5. Raguin DH, Morris GM (1993) Antireflection structured surfaces for the infrared spectral region. Appl Opt 32(7):1154–1167

Chapter 3
Silicon-Based Integrated Diffractive Optical Neural Network Chip

Abstract This chapter will design silicon-based integrated DONN chips for different classification tasks based on the silicon-based integrated DONN model introduced in Chap. 2. Firstly, the structure parameters of the silicon-based integrated DONN are trained separately according to specific classification tasks. Then, the designed silicon-based integrated DONNs are verified through simulation using the optical simulation software 2.5D FDTD. Subsequently, silicon-based integrated DONN chips are fabricated and tested. Additionally, this chapter analyzes the possible system errors and their sources that may arise during chip fabrication, packaging, and experimental testing. The specific chapter arrangements are as follows: Sect. 3.1 will introduce the structure design and simulation verification of the binary classification silicon-based integrated DONN chip. Section 3.2 will cover the structure design, chip fabrication, experimental testing, and system error compensation scheme of the ternary classification silicon-based integrated DONN chip. Section 3.3 will introduce the structure design, chip fabrication, experimental testing, system error compensation, and test result analysis of the ten classification silicon-based integrated DONN chip. Finally, Sect. 3.4 will analyze and discuss the advantages and limitations of the silicon-based integrated DONN chip and compare its performance with that of silicon-based integrated ONNs realized by other methods.

3.1 Binary Classification Silicon-Based Integrated DONN Chip

The binary classification silicon-based integrated DONN chip discussed in this section is designed to classify and predict the diagnostic outcomes of coronary heart disease patients. The dataset used in this research is the UCI Heart Disease Dataset [1], which contains 242 training and 61 testing samples. Each sample has 13 input features and one output feature. The input features represent various medical examination indicators of the patients, while the output feature represents the diagnostic result. If the output feature category is 0, the patient is classified as a non-coronary heart disease patient; if the output feature category is 1, the patient is classified as

© Tsinghua University Press 2026

T. Fu, *Research on Silicon-Based Integrated Diffractive Optical Neural Network*, Springer Theses, https://doi.org/10.1007/978-981-95-3787-7_3

a coronary heart disease patient. The input features are mapped onto the phase of light, which is achieved by heating the phase shifter to change the effective refractive index of the silicon waveguide below the phase shifter, causing the light to produce specific phase differences when passing through the corresponding silicon waveguide, thereby loading the input signal features onto the phase of light and serving as the input signal of the silicon-based integrated DONN.

3.1.1 Design of Structures

The silicon-based integrated DONN structure designed for the binary classification task of the UCI Heart Disease Dataset is shown in Fig. 3.1 (a silicon-based integrated DONN structure with three hidden layers). The input consists of 13 wedge-shaped waveguides, and the output is detected by two monitors that measure the optical power, with the magnitude of the optical power representing the classification result. Each phase weight parameter on the hidden layers of the silicon-based integrated DONN is mapped by an SSSDG structure composed of three identical SSSDs. Each hidden layer in this silicon-based integrated DONN contains 70 distinct phase weight parameters.

During the validation of the binary classification task, this research work trained a total of 5 silicon-based integrated DONNs, and the changes in the loss function (Loss) and blind test accuracy on the test set during the training process are shown in Fig. 3.2a–e. According to the analysis of the training process, it can be seen that the Loss of the silicon-based integrated DONN with different numbers of hidden layers converges and stabilizes, thus indicating that the DONN structure's parameter optimization method is effective. From Fig. 3.2f, it is not difficult to find that the blind test accuracy of the silicon-based integrated DONN does not show a significant

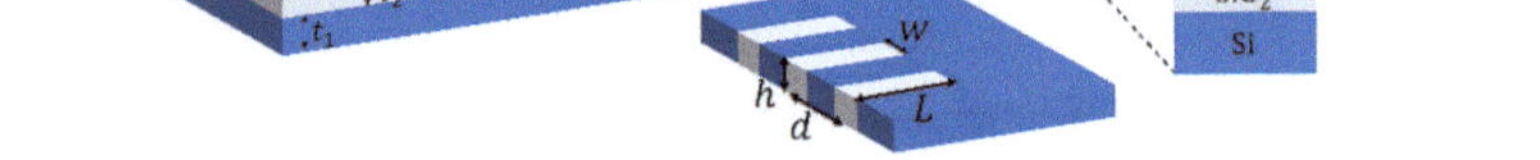

Fig. 3.1 Schematic diagram of silicon-based integrated DONN structure [2]. In this structure, the thickness of the waveguide layer is 220 nm, and the width of the slab waveguide is 105 μm; the distance between the input interface and the hidden layer, the distance between two hidden layers, and the distance between the last hidden layer and the output interface are all 300 μm; the SSSDG consists of three identical SSSDs, and the period of the SSSD is 0.5 μm, while the period of the SSSDG is 1.5 μm

improvement when the number of hidden layers is greater than 3, and the reason for this will be discussed in the conclusion of this chapter.

In this research, the key parameter for neural network training is the phase weight parameter. The following subsection will verify the performance of silicon-based integrated DONNs with 1 hidden layer (DONN-1), 2 hidden layers (DONN-2), and

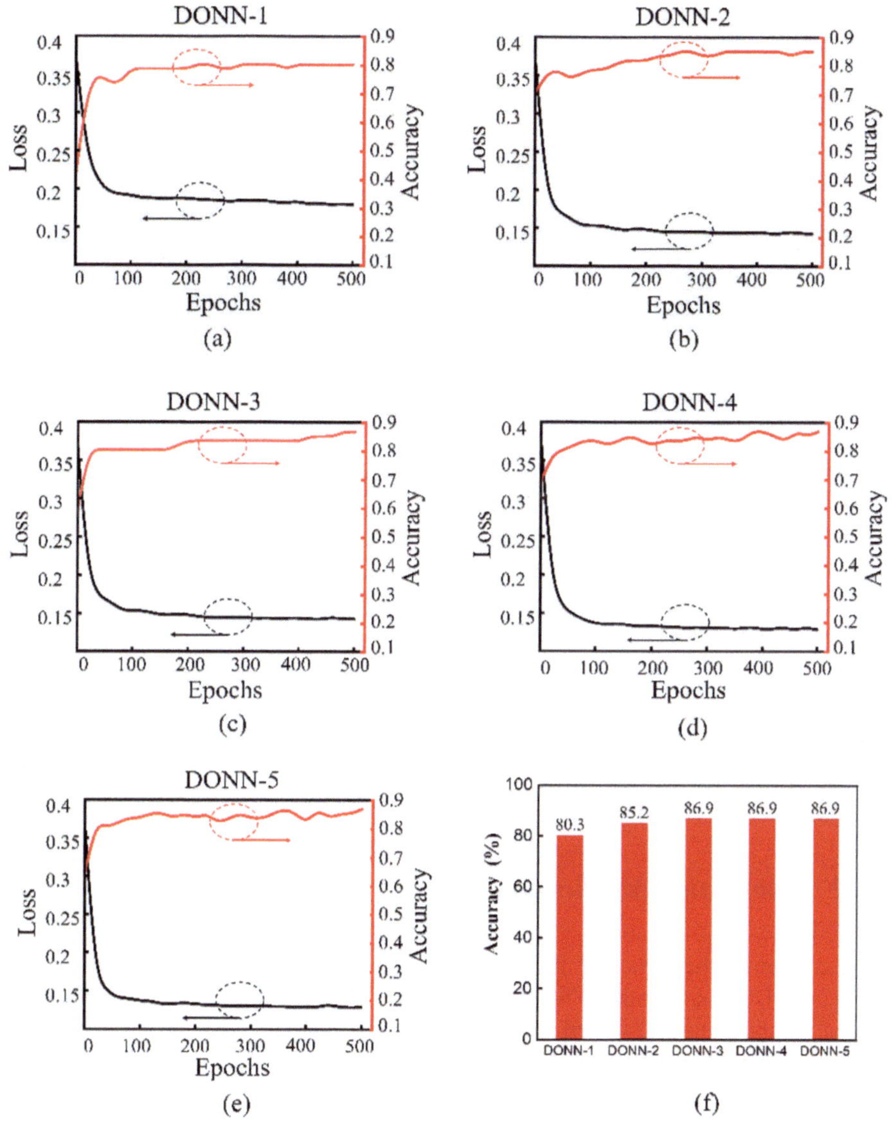

Fig. 3.2 The training process and blind test results of various silicon-based integrated DONNs [2]. Here, DONN-m ($m = 1, 2, 3, 4, 5$) represents that the silicon-based integrated DONN with m hidden layers

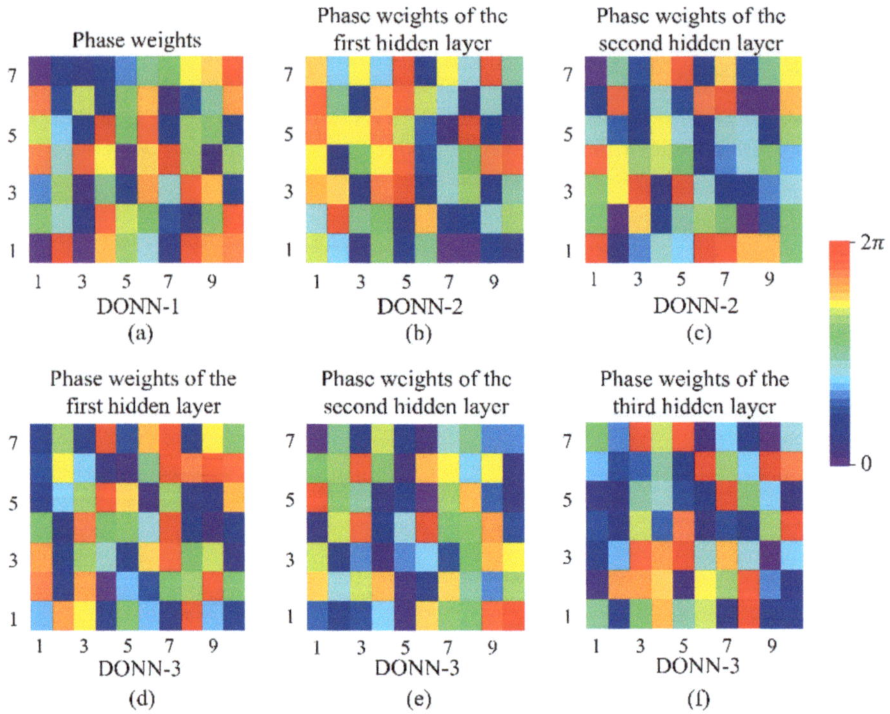

Fig. 3.3 Phase weights distribution on each hidden layer of the silicon-based integrated DONN [2]. Here, DONN-m ($m = 1, 2, 3, 4, 5$) represents a silicon-based integrated DONN with m hidden layers

3 hidden layers (DONN-3) through simulation. Figure 3.3 shows the phase weight parameters on each hidden layer of the three silicon-based integrated DONNs, with 10 pixels per row and 7 rows. Each pixel corresponds to each phase weight on the hidden layer.

3.1.2 *Simulation Verification*

After obtaining the structure parameters of each silicon-based integrated DONN (DONN-1, DONN-2, and DONN-3) through pre-training on the computer, the phase weight mapping model studied in Chap. 2 is used to map the pre-trained parameters onto the physical structure of the DONN chips. The performance of each silicon-based integrated DONN is then verified through simulation using 2.5D FDTD software, with the simulation results shown in Fig. 3.4. From the simulation structure, it can be seen that the designed SSSDG array alters the propagation path of light in the silicon slab waveguide and focuses it on a specific target direction. In the simulation, the monitors are named D_1 and D_2, which have a width of 8 μm, and the center-to-center spacing between them is 40 μm. When the optical power detected by monitor D_1 is greater than that detected by monitor D_2, the prediction is that the patient has coronary heart disease; conversely, when the optical power detected by D_1 is less than that detected by D_2, the prediction is that the patient does not have coronary heart disease. The red dashed lines in Fig. 3.4 indicate the locations of the hidden layers, i.e., the locations of the SSSDG arrays.

To further validate the effectiveness of the SSSDG array design, a set of phase weight parameters was randomly generated in the research. These randomly generated parameters were substituted for the original phase values in the silicon-based integrated DONN-2. The same input signals were fed into the modified silicon-based integrated DONN-2 with the replaced randomly generated phase weight parameters

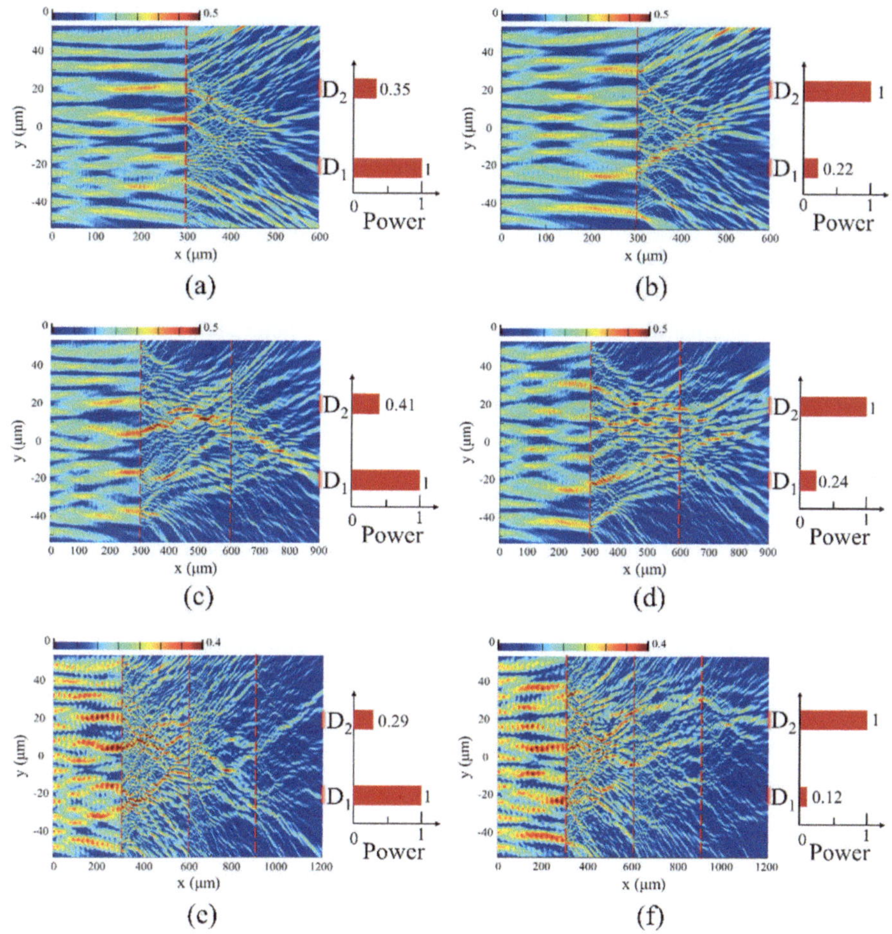

Fig. 3.4 Silicon-based integrated DONN simulation results [2]. **a** and **b** are the 2.5D FDTD simulation results of DONN-1; **c** and **d** are the 2.5D FDTD simulation results of DONN-2; **e** and **f** are the 2.5D FDTD simulation results of DONN-3

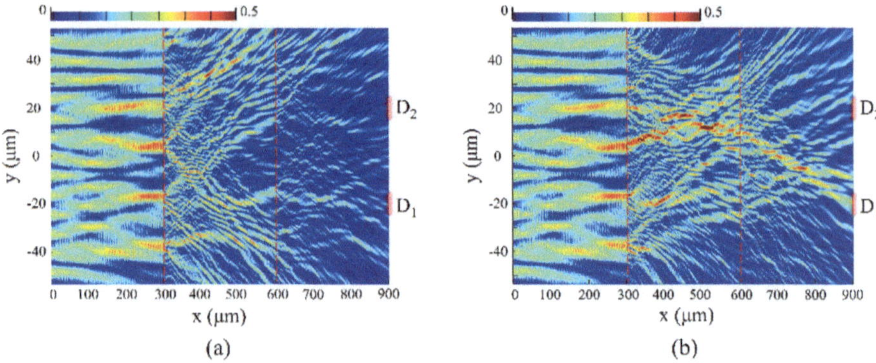

Fig. 3.5 Comparison of simulation results for silicon-based integrated DONN-2 [2]. **a** The phase weight values are randomly generated; **b** The phase weight values are obtained through optimization

for simulation calculations. The results were compared with the original simulation results of the silicon-based integrated DONN-2 as shown in Fig. 3.5 (where the red dashed lines indicate the locations of the hidden layers, i.e., the locations of the SSSDG arrays). In Fig. 3.5a, the randomly generated phase weight parameters result in random control of the optical field, meaning that the optical field distribution after propagation through the SSSDG array is irregular. In contrast, in Fig. 3.5b, the propagation direction of light after passing through the SSSDG array is evident, with the light converging towards monitor D_1, indicating a significant regulatory effect of the phase weight parameters on the light. Therefore, the simulation comparison results in Fig. 3.5 further demonstrate that the phase weight parameters obtained through pre-training are effective in the inference process of the silicon-based integrated DONN.

Next, the pre-trained silicon-based integrated DONN-1, DONN-2, and DONN-3 were used to classify and predict the test set of the UCI Heart Disease Dataset. In this test, classification predictions on the test set were performed using both the silicon-based integrated DONN model and the 2.5D FDTD software. The confusion matrices of the prediction results are depicted in Fig. 3.6. The specific prediction accuracies and the consistency of the prediction results from the two methods are presented in Table 3.1.

According to Table 3.1, in the same silicon-based integrated DONN network, the prediction results based on 2.5D FDTD are lower than those based on the silicon-based integrated DONN model. Moreover, as the number of hidden layers increases, the matching score between the two sets of computing results also decreases. The reason is that the phase weight mapping model has errors under approximate conditions, and these errors accumulate as the number of hidden layers increases.

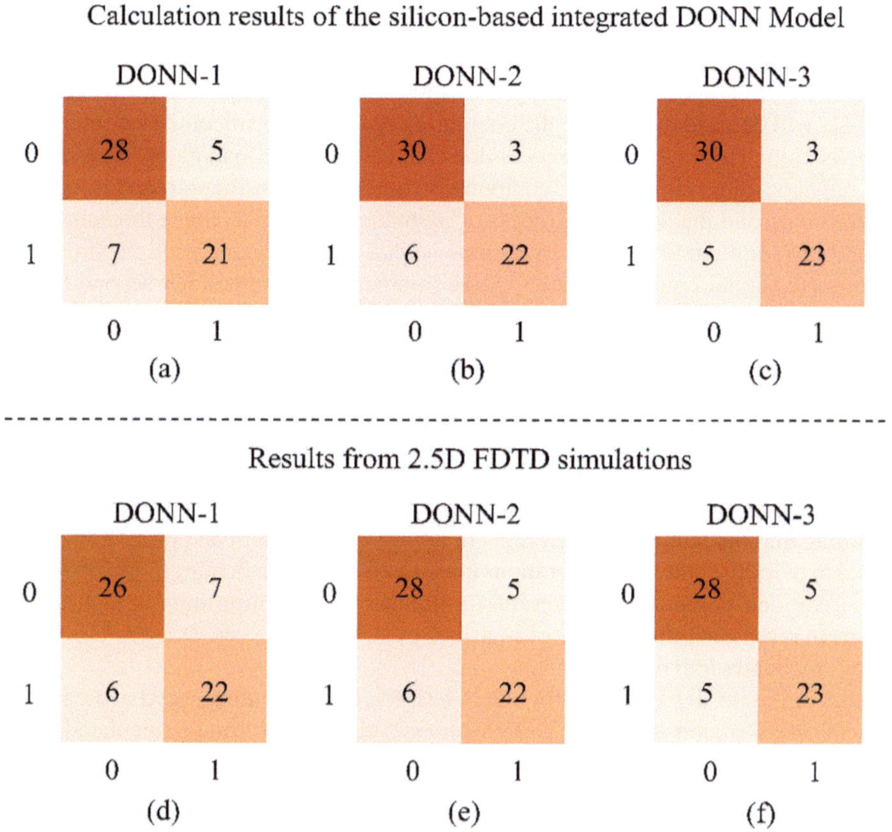

Fig. 3.6 Confusion matrix for the simulation results of the silicon-based integrated DONN test set [2]. Among them, **a–c** show the simulation results of the silicon-based integrated DONN model, while **d–f** show the simulation results of the 2.5D FDTD software

Table 3.1 Comparison of test set predictions between two models [2]

DONN-*m*	Accuracy of the integrated DONN model (%)	Accuracy of the 2.5D FDTD simulation (%)	The matching score between the DONN model and the 2.5D FDTD simulation results (%)
DONN-1	83.3	78.7	95.1
DONN-2	85.2	82.0	93.4
DONN-3	86.9	83.6	91.8

3.1.3 Performance Analysis

Once the structure parameters of the silicon-based integrated DONN chip are pre-trained, the entire process of executing a task will rely on the interference and diffraction of light. The whole computing process will be carried out in a passive manner, where the working state of the computing units does not require an external energy supply. Therefore, the silicon-based integrated DONN chip has the advantages of high computing speed, low latency, and low power consumption. In this section, Eq. 3.1 will be used to evaluate the computing speed of the silicon-based integrated DONN chip. This evaluation process does not consider the loading of input signals but only considers the inference computation process of the silicon-based integrated DONN chip and the detection process of output optical power. Since the number of operations required to perform an N-dimensional matrix–vector multiplication on a traditional digital computer is $O(N^2)$, the number of operations per second (OPS) that matched with the optical network can be calculated using Eq. 3.1 [3].

$$R = 2m \times N^2 \times D_{\mathrm{bw}} \text{ OPS} \tag{3.1}$$

where R represents the number of operations per second, the equation is multiplied by a coefficient of 2 to indicate that a single operation process in an optical network includes both "multiplication" and "addition" operations (a total of two operations), meaning that the number of operations in a single operation of an optical network is equal to twice the number of operations performed in a single floating-point operation on a traditional computer; N represents the number of computing units on each hidden layer; m represents the number of matrices of size $N \times N$; and D_{bw} represents the detection bandwidth of the detector.

Assuming the detector's bandwidth is 40 GHz, the computing speed of the binary classification silicon-integrated DONN ($m = 2$) chip, which can be calculated using Eq. 3.1, is 7.84×10^{14} OPS (i.e., 7.84×10^{14} operations per second).

For the time delay calculation of silicon-based integrated DONN, the current calculation only considers the time required for light to pass through the entire on-chip system, and Eq. 3.2 is the delay calculation equation as shown below.

$$\begin{aligned} T_L = D_{\mathrm{nw}} \times \left(\frac{c_0}{n_{g1}}\right)^{-1} + M \times D_m \times \left(\frac{c_0}{n_{g2}}\right)^{-1} \\ + \left(D_{\mathrm{wf}} + (M-1) \times D_p + D_f\right) \times \left(\frac{c_0}{n_{g3}}\right)^{-1} \end{aligned} \tag{3.2}$$

where T_{L} represents the time that light travels through the silicon waveguide between the input and output ends; c_0 represents the speed of light in vacuum; D_{nw} represents the distance from the narrow waveguide to the wide waveguide, i.e., the length of the wedge waveguide; n_{g1} represents the group refractive index of the narrow waveguide (a silicon waveguide with a width of 450 nm and a height of 220 nm); D_{m} represents the length of the longest SSSDG array in the chip; M represents the total number of SSSDG arrays; n_{g2} represents the group refractive index of the SSSDG region; D_{wf} represents the distance from the input interface to the first hidden layer; D_{p} represents the distance between adjacent hidden layers; D_{f} represents the distance from the last hidden layer to the output interface; n_{g3} represents the group refractive index of the silicon-based slab waveguide. Here, we use the maximum group refractive index,

$\max(n_{g1}, n_{g2}, n_{g3})$, to approximately estimate the delay of light passing through the binary classification silicon-integrated DONN-1 chip. The maximum delay will not exceed 11.11 ps.

The prediction accuracy of the designed silicon-based integrated DONN-3 chip on the test set of the UCI Heart Disease Dataset is 86.9%, while the prediction accuracy achieved by other methods is distributed between 80.0% and 89.0% [4–9]. The prediction results of the silicon-based integrated DONN-3 chip are nearly consistent with the classification results achieved by other methods.

3.2 Ternary Classification Silicon-Based Integrated DONN Chip

In Sect. 3.1, the silicon-based integrated DONN structure parameters were designed for the binary classification task using the UCI Heart Disease Dataset. The performance of silicon-based integrated DONNs with different numbers of hidden layers was verified through simulations, which further demonstrated the effectiveness of the silicon-based integrated DONN model. Therefore, this section will design the silicon-based integrated DONN structure parameters for the ternary classification task using the IRIS Plants Dataset [10] and validate the designed silicon-based integrated DONN's performance by means of simulation and experimentation.

The IRIS Plants Dataset comprises 150 samples in total, including 120 training samples and 30 testing samples. Each sample has four input features and three output features. The input features correspond to the structure dimensions of the iris flower, namely sepal length, sepal width, petal length, and petal width. The output features represent the species of the iris flower, which include Setosa, Versicolor, and Virginica. Specifically, the input features are mapped onto the phase of light and then loaded onto the optical waveguide via thermo-optic phase shifters to serve as the input for the silicon-based integrated DONN. The output features are represented by the values 0, 1, or 2, corresponding to the three iris species: Setosa (0), Versicolor (1), and Virginica (2). In the structure design, these categories are represented by three distinct monitors, D_1, D_2, and D_3, which correspond to Setosa (0), Versicolor (1), and Virginica (2), respectively.

3.2.1 Design of Structures

The structure design of the ternary classification silicon-based integrated DONN chip is the same as the method described in Sect. 2.3. First, the structure parameters of DONN for the classification task of the IRIS Plants Dataset are obtained through iterative training using the silicon-based integrated DONN model. The key parameters optimized during this training process are the phase weights, while other parameters are fixed before the entire network training. These parameters, which are fixed before the training of the entire silicon-based integrated DONN, are referred to as hyperparameters. It should be noted that in the actual fabricated silicon-based integrated DONN chip, the monitors are realized through wedge-shaped waveguides transitioning to single-mode waveguides, and then the light is propagated from the single-mode waveguides to vertical coupling gratings to be coupled out of the chip. The optical power coupled out of the chip by the vertical coupling grating originates mainly from the TE_0 mode. Therefore, during the training process of the silicon-based integrated DONN chip's structure parameters, it is necessary to consider the mode selection of light received in the monitor region. In this training process, the mode overlap integral is considered to filter the TE_0 mode light. The specific implementation can be found in Eqs. 3.3, 3.4, and 3.5.

$$a_0 = \frac{(\boldsymbol{E}_y)^{\mathrm{T}} \cdot \overline{\boldsymbol{H}}_{0z}}{(\boldsymbol{E}_{0y})^{\mathrm{T}} \cdot \overline{\boldsymbol{H}}_{0z}} \tag{3.3}$$

$$b_0 = \frac{(\overline{\boldsymbol{E}}_{0y})^{\mathrm{T}} \cdot \boldsymbol{H}_z}{(\overline{\boldsymbol{E}}_{0y})^{\mathrm{T}} \cdot \boldsymbol{H}_{0z}} \tag{3.4}$$

$$P_{\mathrm{TE}_0} = a_0 \cdot b_0 \tag{3.5}$$

Specifically, $\boldsymbol{E}_y$ and $\boldsymbol{H}_z$ represent the electric and magnetic fields in the region to be detected by the monitor, respectively; $\boldsymbol{E}_{0y}$ and $\overline{\boldsymbol{E}}_{0y}$ are the electric field and its conjugate of the output single-mode waveguide; $\boldsymbol{H}_{0z}$ and $\overline{\boldsymbol{H}}_{0z}$ are the magnetic field and its conjugate of the output single-mode waveguide, and P_{TE_0} represents the optical power of the TE_0 mode. After considering the mode selection, the optical power received by the monitor in the silicon-based integrated DONN chip structure parameter training process will correspond to the actual physical process of the light passing through the output coupler and being received by the detectors. Thus, it will not affect the prediction results of the classification task.

In this section, a total of five silicon-based integrated DONNs were pre-trained, and the performance of silicon-based integrated DONN-I1 and DONN-I3 will be experimentally verified. The structure diagram of the silicon-based integrated DONN-I1 chip is shown in Fig. 3.7. The structures of the other DONN chips are similar to that of the silicon-based integrated DONN-I1 and will not be repeated here. In the silicon-based integrated DONN structures designed in this research, each hidden layer has 186 phase weight parameters, and each phase weight parameter is physically implemented by an SSSDG with a period of 1.5 μm. Additionally, the distance from the input interface to the first hidden layer and from the last hidden layer to the output interface is 250 μm, and the distance between all adjacent hidden layers is also 250 μm. Three monitors are arranged at the output interface, each with a width of 8 μm and a center-to-center spacing of 70 μm between adjacent monitors.

This section presents the loss function changes during the silicon-based integrated DONN-I1 and DONN-I3 training process, as shown in Fig. 3.8a, b, respectively.

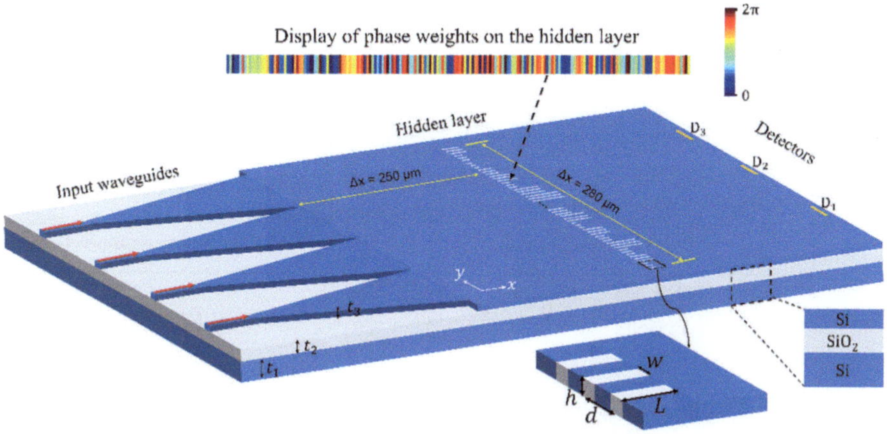

Fig. 3.7 Schematic diagram of silicon-based integrated DONN chip [11]. In this figure, the thickness of the waveguide layer is 220 nm, and the width of the silicon-based slab waveguide is 300 μm. The distance between the input interface and the hidden layer and the distance between the hidden layer and the output interface are both 250 μm. The SSSDG consists of three identical SSSDs. The period of the SSSD is 0.5 μm, so the period of the SSSDG is 1.5 μm

Figures 3.8c, d are the confusion matrices for the test set of the silicon-based integrated DONN-I1 and DONN-I3, respectively. A confusion matrix, also known as an error matrix, is often used in artificial intelligence applications as a visualization tool to illustrate the prediction performance of a neural network for classification tasks. The neural network's prediction results from the confusion matrix among different classes can be observed. For example, the entry in the third row and second column of Fig. 3.8c indicates that the silicon-based integrated DONN incorrectly predicted the third class of iris flower (Virginia) as the second class of iris flower (Versicolor). The entry in the second row and first column of Fig. 3.8d indicates that the DONN incorrectly predicted the second class of iris flower (Versicolor) as the first class of iris flower (Setosa). Meanwhile, the entries in the first row and first column of Fig. 3.8c, d indicate that the silicon-based integrated DONN correctly predicted all instances of the first class of iris flowers (Setosa).

The optimal phase weight parameters for each silicon-based integrated DONN chip can be obtained through iterative optimization using the silicon-based integrated DONN model. Although this section only presents the training process and prediction results on the test set for the silicon-based integrated DONN-I1 and DONN-I3, structure parameter training has also been conducted for other DONNs with varying numbers of hidden layers in this work. Table 3.2 illustrates the blind test results on the IRIS test set after training the parameters for different silicon-based integrated DONNs. Under this structure, the increase in test set accuracy is not significant as the number of hidden layers increases. Therefore, when dealing with specific tasks,

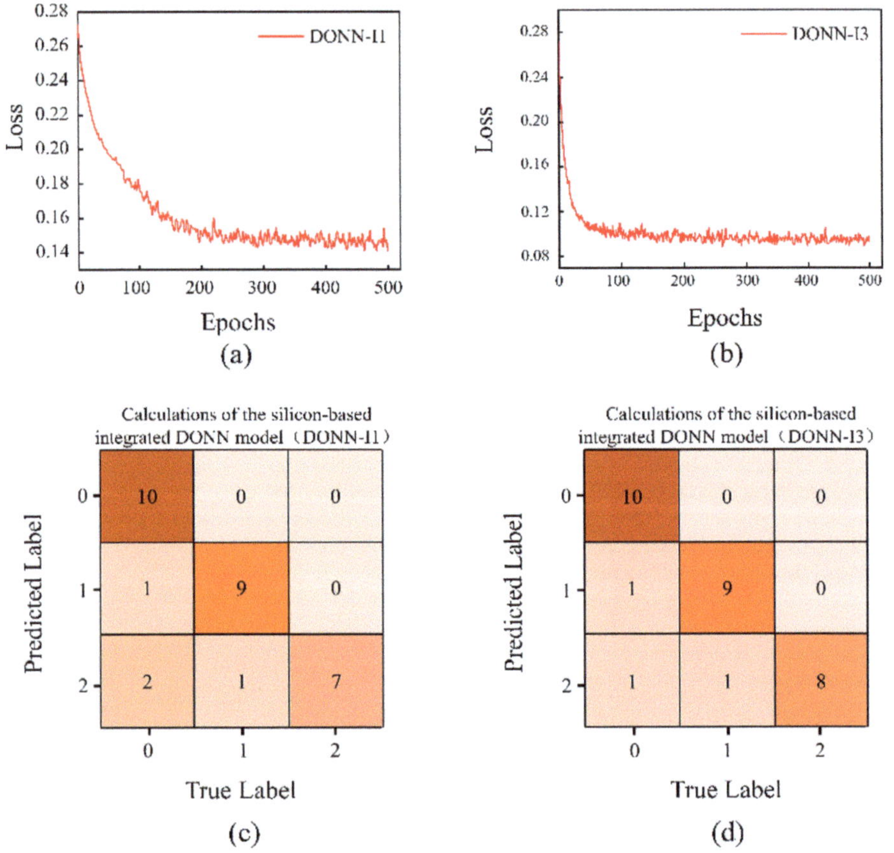

Fig. 3.8 The variation of loss during the training process of the silicon-based integrated DONN and the confusion matrix for the IRIS test set. **a**, **b** show the changes in the loss function for the silicon-based integrated DONN-I1 and DONN-I3, respectively; **c**, **d** are the confusion matrices for the IRIS test set corresponding to the silicon-based integrated DONN-I1 and DONN-I3, respectively

it is necessary to comprehensively consider the network scale cost of the silicon-based integrated DONN and the specific task requirements to determine the specific structure parameters.

Table 3.2 IRIS plants dataset test set prediction results

DONN-*m*	Accuracy of the integrated DONN model (%)
DONN-I1	86.7
DONN-I2	86.7
DONN-I3	90.0
DONN-I4	90.0
DONN-I5	90.0

Based on the computational results of the silicon-based integrated DONN model, the DONN-I1 and DONN-I3 were selected for experimental verification in this study. Therefore, Fig. 3.9 presents the phase weight parameters obtained after the optimization process of these two silicon-based integrated DONNs.

Finally, to validate the effectiveness of the silicon-based integrated DONNs designed for the classification task of the IRIS Plants Dataset, this research conducted simulation comparison experiments on the two structures, DONN-I1 and DONN-I3, and the specific results are presented in Fig. 3.10. The comparison results indicate that when the phase weight parameters are optimized through training, the performance of the silicon-based integrated DONN-I1 and DONN-I3 (labeled as index number 1 in the figure) is significantly higher than that of other silicon-based integrated DONNs with randomly assigned phase weight parameters. This conclusion is consistent with the validation results in Sect. 3.1.2 (Fig. 3.5), thereby further demonstrating the effectiveness of the optimized phase weight parameters.

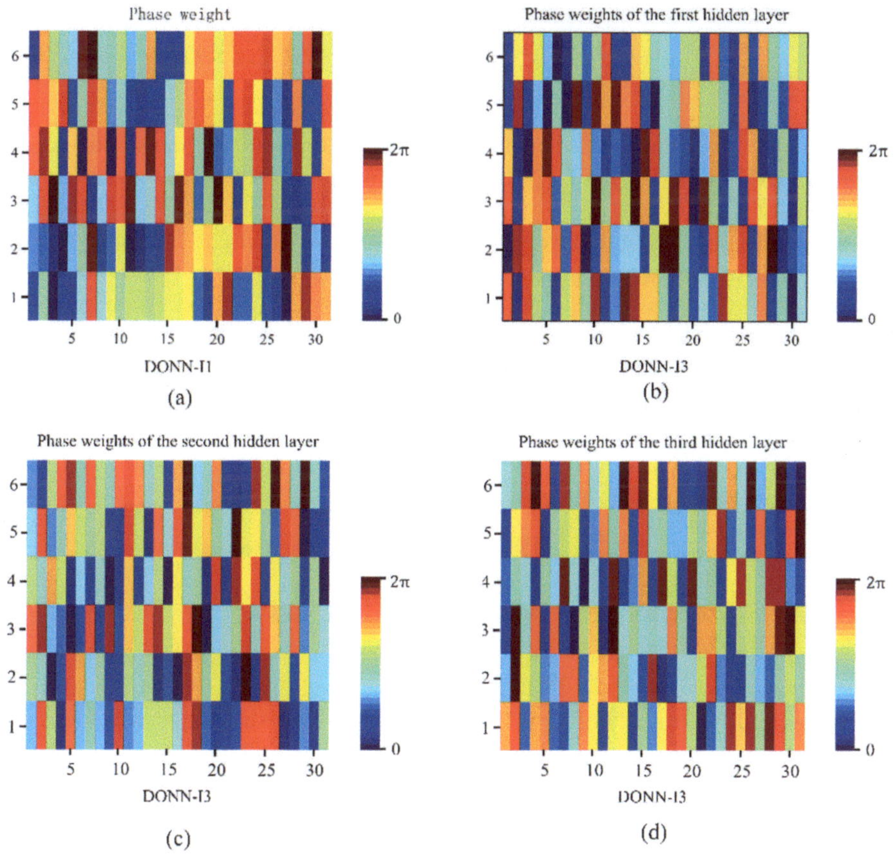

Fig. 3.9 Phase weight distribution on the hidden layers of the silicon-based integrated DONNs. DONN-I1 and DONN-I3 represent the silicon-integrated DONN with 1 and 3 hidden layers, respectively

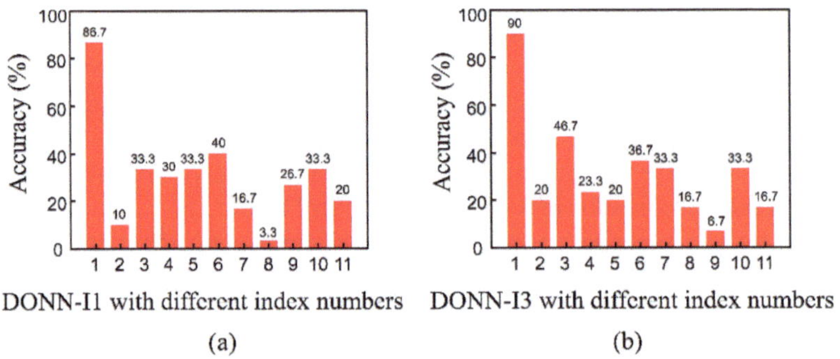

Fig. 3.10 Validation of the effectiveness of phase weight parameter optimization for silicon-based integrated DONNs [11]. The silicon-based integrated DONN-I1 and DONN-I3 labeled as index number 1 have phase weight parameters obtained through optimized training. The phase weight parameters of other numbered silicon-based integrated DONN-I1 and DONN-I3 are randomly assigned. The vertical axis represents the blind test accuracy of each silicon-based integrated DONN on the test set of the IRIS Plants Dataset

3.2.2 Simulation Verification

This subsection will introduce the simulation validation results of the silicon-based integrated DONN-I1. Figure 3.11 shows the simulation prediction results for the three types of iris flowers.

Specifically, Fig. 3.11a, b have input features corresponding to Setosa, resulting in a higher detection power at monitor D_1 on the output interface than D_2 and D_3, indicating a correct prediction of the iris flower. Figure 3.11c, d have input features related to Versicolor, with the detection power at monitor D_2 on the output interface being higher than D_1 and D_3, indicating a correct prediction for Versicolor. Similarly, Fig. 3.11e, f have input features corresponding to Virginica, with the detection power at monitor D_3 on the output interface being higher than D_1 and D_2, thus accurately predicting the class of Virginica. Additionally, in Fig. 3.11, the unit "a.u." for normalized intensity is an arbitrary unit.

3.2.3 Experimental Tests and Results Analysis

In the preceding chapters, the DONNs designed based on the DONN model have undergone multiple rounds of simulation validation. This subsection will focus on the fabrication of silicon-based integrated DONN-I1 and DONN-I3 chips for the classification task of the IRIS Plants Dataset, as well as the completion of chip packaging and experimental testing. From the beginning of the chip design, this research has thoroughly considered the fabrication conditions and the packaging requirements after chip fabrication. Only after extensive research and communication did we commence the design of the silicon-based integrated DONN chip for the

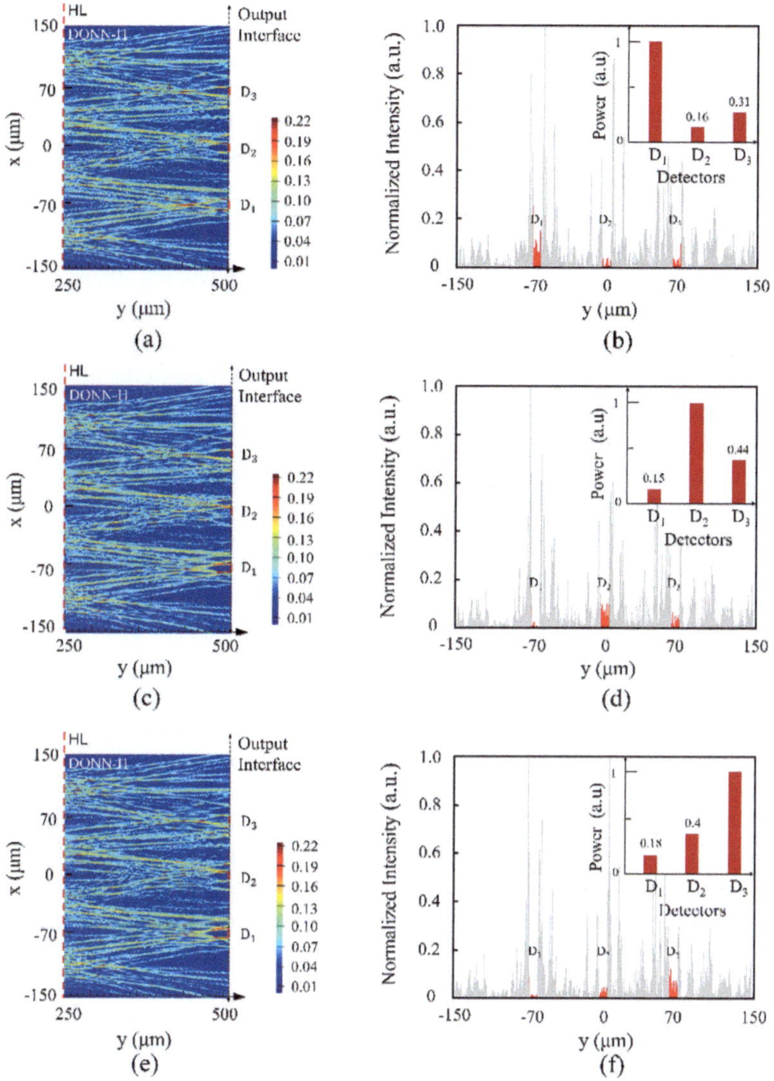

Fig. 3.11 Simulation prediction results of the silicon-based integrated DONN-I1 for different types of iris flowers [11]. Here, D_1, D_2, and D_3 represent the prediction records for different types of iris flowers, namely Setosa, Versicolor, and Virginica. The red dashed lines in the figure indicate the positions of the hidden layers (HL). For **a** and **b**, the simulation result for prediction of Setosa, where the power at D_1 should be greater than that at D_2 and D_3, indicates a correct prediction result. The interpretation of **c** and **d**, as well as **e** and **f**, is the same as that of (**a**) and (**b**)

ternary classification task. The minimum feature size in the silicon-based integrated DONN chip structure design is 200 nm, the critical dimension to meet the fabrication requirements. The chip fabrication includes two types of networks: silicon-based integrated DONNs with one hidden layer and three hidden layers, respectively. Here, we will elaborate on the key aspects of the chip fabrication process. The entire silicon-based integrated DONN chip is fabricated on an SOI platform. The buried oxide layer (lower cladding) on the Si substrate (100) has a thickness of 3 μm, and the silicon waveguide layer above the buried oxide layer has a thickness of 220 nm. All structure patterns of the silicon-based integrated DONN are realized on the silicon waveguide layer through electron-beam lithography, reactive ion etching, etc. Subsequently, a 2 μm thick SiO_2 upper cladding layer is deposited. Next, thermos-optic phase shifters, metal wires, and metal electrodes are fabricated on the surface of the upper cladding using alignment and photolithography processes. The material for the thermos-optic phase shifters is titanium nitride (TiN) alloy, while aluminum-copper (AlCu) alloy is used for the metal wires and electrodes. Finally, a 2 μm thick SiO_2 protective layer is deposited on the chip surface.

The entire chip fabrication process can be understood as a multi-step patterning process, which involves sequentially realizing the patterns from different layers of the chip layout on the SOI platform, layer by layer. Many tools are available for drawing chip layouts, and K-Layout software was used in this instance. The DONN-I1 and DONN-I3 structures in the GDS file are shown in Fig. 3.12.

After fabrication according to the defined layout structure, the silicon-based integrated DONN-I1 and DONN-I3 chips can be observed under a microscope, as shown in Fig. 3.13. Figure 3.13a shows the microscope image of the silicon-based integrated DONN-I1 chip. Here, we briefly introduce the propagation process of light throughout the chip. First, the laser enters a single-mode waveguide (with a silicon waveguide width of 450 nm and height of 220 nm) through an input coupler (vertical coupling grating). Then, the light is split into four beams via a two-stage MMI (each propagating in a single-mode waveguide) and travels through wedge-shaped waveguides to reach the input interface of the silicon slab waveguide (i.e., the interface between the wedge-shaped waveguide and the slab waveguide). Each of the four single-mode waveguides is equipped with a thermo-optic phase shifter, which is connected to a metal electrode via metal wiring. The function of these four thermo-optic phase shifters is to load the input signals. Specifically, different voltages are applied to the thermo-optic phase shifters. Under different voltages, the phase shifters generate varying thermal effects, which alter the effective refractive index of the single-mode waveguide beneath the corresponding phase shifter. This, in turn, causes different phase delays in the light propagating through that section of the waveguide, thereby loading the input signal features onto the phase of light. When the light reaches the silicon slab waveguide's input interface, the propagating light's phase has already carried the feature information of the input signal. As the phase-encoded light continues propagating through the SSSDG array, each SSSDG imposes a specific phase modulation on the propagating light. This results in a particular interference pattern at the output interface of the silicon slab waveguide, where the light is coupled out of the chip through an output coupler in a specific region (the monitor placement

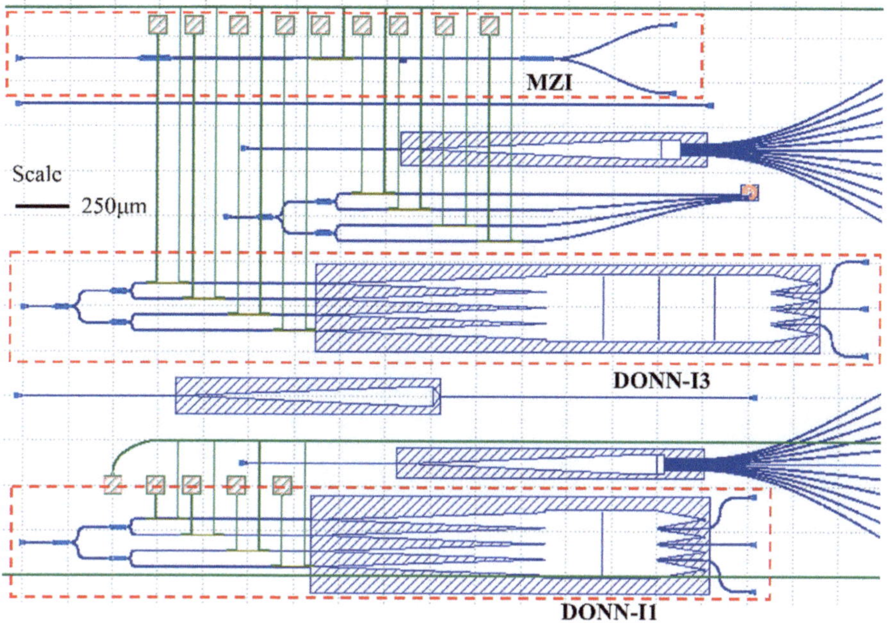

Fig. 3.12 Silicon-based integrated DONN chip layout design for the IRIS Plants Dataset. The graphical structures within the dashed box in the figure represent the layouts of the silicon-integrated DONN-I1, DONN-I3, and MZI

area). This entire process constitutes a complete computation cycle. The optical power detected by the three monitors can be used to determine which category of iris flower the input signal features belong to, thereby making a classification prediction. Additionally, in Fig. 3.13a, three larger wedge-shaped waveguides are connected to an optical attenuator at the output interface. The purpose of these attenuators is to prevent light reflection at the output interface from returning to the SSSDG array, which could otherwise interfere with the DONN system. Figure 3.13b shows a bare die after dicing following the chip fabrication process, with dimensions (length × width) of approximately 8 mm × 3 mm. Figure 3.13c illustrates the testing of the bare die to inspect whether the optical paths between the input and output couplers have been damaged by scratches or other defects. This step is typically completed before chip packaging, as packaging costs are often high, and it is a crucial verification step that must be performed before proceeding with packaging. Figure 3.13d is a microscope image of the silicon-based integrated DONN-I3 chip, whose computational process is identical to that of the aforementioned silicon-based integrated DONN-I1 chip.

After the bare die testing is completed, the bare die needs to be packaged to facilitate the performance testing of the silicon-based integrated DONN chips in subsequent steps. The key packaging steps include the following two points: First, wire bonding involves connecting the metal electrodes on the chip to the electrodes

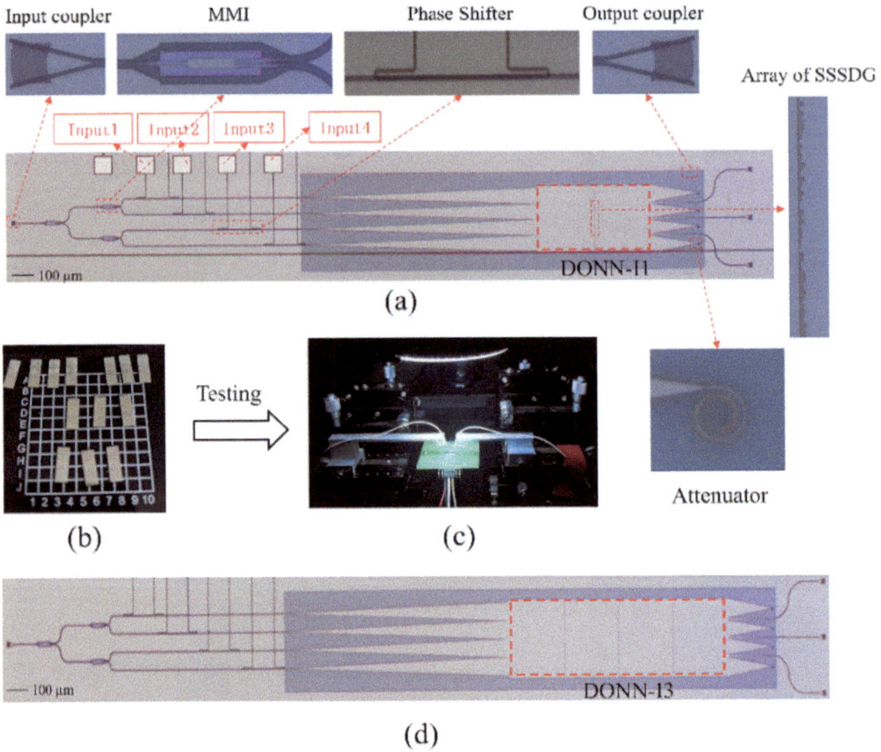

Fig. 3.13 Optical microscope structure and bare die testing process of the silicon-based integrated chips. **a** Microscope image of the silicon-based integrated DONN-I1 chip; **b** Bare dies after dicing; **c** Demonstration of bare die testing; **d** Microscope image of the silicon-based integrated DONN-I3 chip

on the PCB board using metal wires. The green substrate in Fig. 3.13c is the PCB board. The second step is optical packaging, which involves optimizing the alignment positions of the optical fibers with the input and output grating couplers to maximize the optical power entering and exiting the chip. Subsequently, the optical fibers are fixed to the chip's input and output grating couplers using FA (fiber alignment) techniques. This ensures the stability of the entire system during the subsequent chip performance testing, facilitating the smooth progress of the experiments. The packaging process of the silicon-based integrated DONN-I1 and DONN-I3 chips is shown in Fig. 3.14. After packaging, the dimensions (length × width × height) of the packaged silicon-based integrated DONN chip are approximately 59 mm × 54 mm × 28 mm.

Once the overall packaging of the chip is completed, the experimental testing of the silicon-based integrated DONN chip can be carried out. As shown in Fig. 3.15, the experimental system setup includes instruments such as a laser, a polarization controller, a voltage source, a driver module, a multi-channel optical power meter,

Fig. 3.14 Illustration of the packaging process of the silicon-based integrated DONN chip. **a** Physical image of the packaged silicon-based integrated DONN chip; **b** Close-up of the chip packaging; **c** Close-up of wire bonding on the chip electrodes, where the two white fixtures are FA (fiber alignment)

communication data lines, and a CPU. The laser provides the light source with a wavelength of 1550 nm. The polarization controller adjusts the laser to a polarization state conducive to grating coupling, enabling more optical power from the laser to be coupled into the on-chip waveguides. The voltage source powers the driver module, which includes a communication device and a voltage amplifier. Its function is to receive signals transmitted from the CPU and, according to the CPU's instructions, apply specific voltages to the electrodes on the PCB board, thereby further loading the corresponding voltages onto the metal electrodes of the chip. The silicon-based integrated DONN chip performs the classification inference task based on the input signal features and the pre-trained hidden layer structures. The multi-channel detector receives the optical power from the output coupling gratings of the chip. After the multi-channel detector collects optical power at each output port of the chip, it transmits the intensity measurement results to the CPU. The CPU then processes this data uniformly and sends the next round of signals, repeating this cycle until the entire classification task is completed.

Once the experimental testing apparatus is set up, the test set of the IRIS Plants Dataset can be experimentally tested. However, before testing, it is necessary to know the voltage-phase curve. This is because the features of the iris flowers are loaded onto the phase of light, and the phase delay is achieved by heating the thermo-optic phase shifter to change the effective refractive index of the corresponding waveguide, thereby realizing the phase delay of the propagating light to load the input signals. Therefore, it is essential to know the quantitative relationship between the voltage and

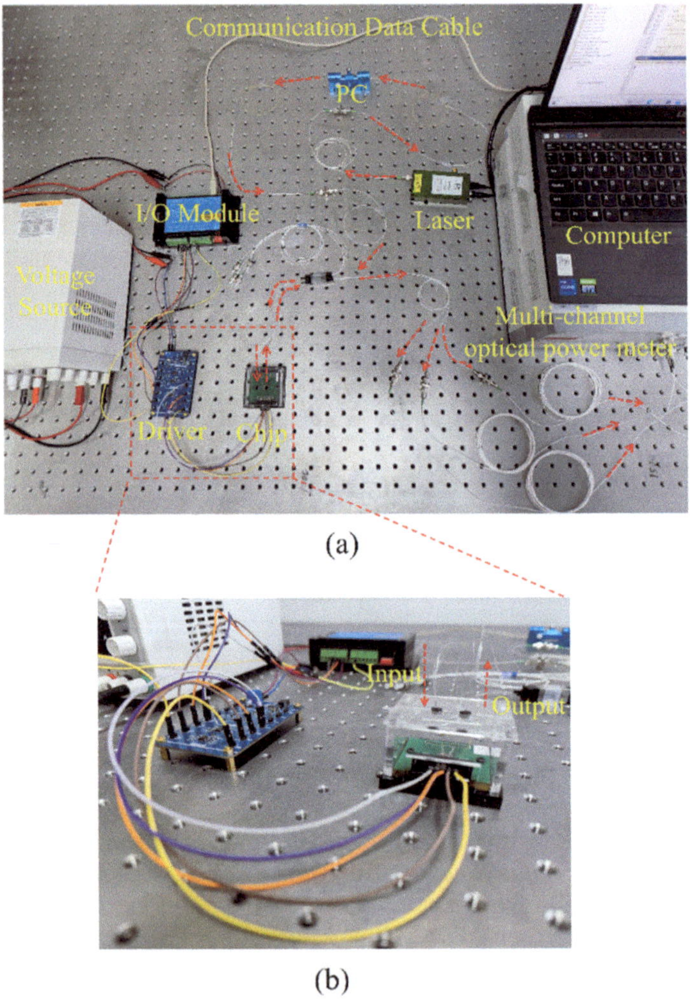

Fig. 3.15 Testing apparatus of the silicon-based integrated DONN chip. **a** Experimental testing system setup for the silicon-based integrated DONN chip; **b** Close-up of a part of the testing system

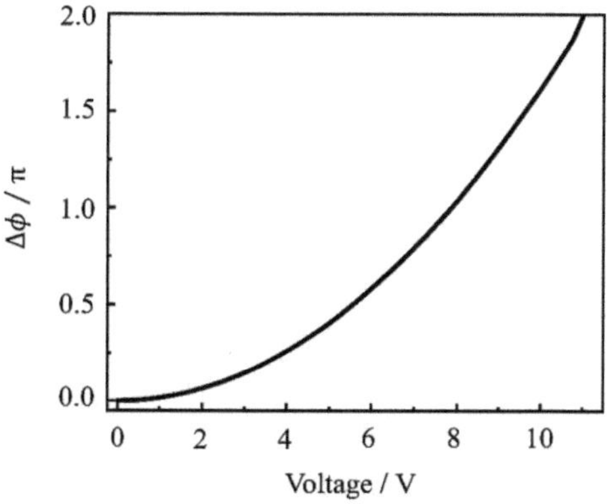

Fig. 3.16 Voltage-phase curve

the phase shift it can cause. The voltage-phase curve can be calculated by sweeping the voltage on the thermo-optic phase shifter in one arm of an MZI, as shown in Fig. 3.16.

Based on the voltage-phase curve shown in Fig. 3.16, the classification prediction work for the test set of the IRIS Plants Dataset can be carried out. The test results indicate that the prediction accuracies of the silicon-based integrated DONN-I1 and DONN-I3 chips on the test set are 56.7% and 60.0%, respectively. These results differ significantly from the prediction accuracies obtained through simulation, which were 86.7% and 90.0%, respectively. The large discrepancy between the experimental and simulation results is attributed to system errors in the experimental process. The sources of these system errors can be mainly categorized into the following aspects: first, errors arising during the loading of input signals; second, inevitable fabrication errors during the chip process stage; lastly, errors introduced during the packaging process.

Although system errors persist at this stage and have led to suboptimal performance of the silicon-based integrated DONN chips, this occurrence represents a normal phenomenon in the fabrication process. This is attributed to the fact that the performance of most systems is influenced by varying degrees in the absence of error compensation. Therefore, to ensure the good performance of the silicon-based integrated DONN chips, the following Sect. 3.2.4 will begin with an analysis of process errors to discuss the entire experimental system errors and propose an effective error compensation scheme.

3.2.4 System Error Compensation

In Sect. 3.2.3, the experimental setup for the silicon-based integrated DONN chip has been completed, and the preliminary performance testing of the silicon-based integrated DONN chip has also been carried out. However, due to experimental system errors, the prediction accuracies of the silicon-based integrated DONN-I1 and

DONN-I3 chips on the IRIS Plants Dataset (test set) were only 56.7% and 60.0%, respectively. These results differ significantly from the simulation results. Therefore, this section will conduct a detailed analysis of the sources of experimental system errors.

Firstly, assuming that system errors mainly originate from fabrication stage, this subsection will begin with a simple example to analyze the errors introduced during the chip fabricating process. Figure 3.17 shows the structure and logical diagrams of the silicon-based integrated DONN. The matrix operation process of the silicon-based integrated DONN can be represented by a mathematical expression, as shown in Eq. 3.6, where the SSSDG array is a diagonal matrix. According to Eq. 2.14, the length of each SSSDG can be calculated based on the phase weight parameters obtained during optimization training. Ideally, the physical process in Fig. 3.17a can be represented by Eq. 3.6. However, the fabrication of SSSDG involves three steps: First, subwavelength rectangular micro-slot patterns are formed on the waveguide layer using photolithography; Second, rectangular micro-slots are etched in the lithographed areas using reactive ion etching; Finally, SiO_2 material is deposited to fill the etched subwavelength micro-slots. After the fabrication process, the SSSDGs obtained may implement phase weights that differ from the theoretical design, thereby introducing errors into the system. Since the SSSDG consists of subwavelength structures with an etching depth of 220 nm for the rectangular micro-slots, achieving uniform and smooth etching edges is challenging. Additionally, the uniformity of SiO_2 filling at different positions cannot be guaranteed during the filling process. Therefore, it can be concluded that fabrication errors are almost inevitable during the processing. As a result, for the fabricated silicon-based integrated DONN chips, the mathematical matrix representation of the physical process is shown in Eq. 3.7, meaning that each phase weight may have an error term. For the theoretically designed silicon-based integrated DONN chip, the detected optical power can be obtained using Eq. 3.8. However, after the introduction of fabrication errors, the optical power of the silicon-based integrated DONN chip should be calculated using Eq. 3.9. In this way, the output optical power error caused by fabrication process is analyzed in mathematical expressions.

$$\begin{bmatrix} y_1 \\ y_2 \\ y_3 \end{bmatrix} = \begin{bmatrix} \mathbf{W}^{(2)} \end{bmatrix}^{\mathrm{T}}_{3\times m} \begin{bmatrix} e^{j\phi_{11}} & 0 & 0 & \cdots & 0 \\ 0 & e^{j\phi_{22}} & 0 & \cdots & 0 \\ \vdots & \ddots & \ddots & \ddots & \vdots \\ 0 & \cdots & 0 & \ddots & 0 \\ 0 & \cdots & 0 & 0 & e^{j\phi_{mm}} \end{bmatrix} \begin{bmatrix} \mathbf{W}^{(1)} \end{bmatrix}^{\mathrm{T}}_{m\times 4} \begin{bmatrix} x_1 \\ x_2 \\ x_3 \\ x_4 \end{bmatrix} \tag{3.6}$$

$$\begin{bmatrix} y_1^{\mathrm{err}} \\ y_2^{\mathrm{err}} \\ y_3^{\mathrm{err}} \end{bmatrix} = \begin{bmatrix} \mathbf{W}^{(2)} \end{bmatrix}^{T}_{3\times m} \begin{bmatrix} e^{j\left(\phi_{11}+\phi_{11}^{err}\right)} & 0 & 0 & \cdots & 0 \\ 0 & e^{j\left(\phi_{22}+\phi_{22}^{\mathrm{err}}\right)} & 0 & \cdots & 0 \\ \vdots & \ddots & \ddots & \ddots & \vdots \\ 0 & \cdots & 0 & \ddots & 0 \\ 0 & \cdots & 0 & 0 & e^{j\left(\phi_{mm}+\phi_{mm}^{\mathrm{err}}\right)} \end{bmatrix} \begin{bmatrix} \mathbf{W}^{(1)} \end{bmatrix}^{T}_{m\times 4} \begin{bmatrix} x_1 \\ x_2 \\ x_3 \\ x_4 \end{bmatrix} \tag{3.7}$$

$$\begin{bmatrix} P_1 \\ P_2 \\ P_3 \end{bmatrix} = \begin{bmatrix} |y_1|^2 \\ |y_2|^2 \\ |y_3|^2 \end{bmatrix} \quad (3.8)$$

$$\begin{bmatrix} P_1^{\mathrm{err}} \\ P_2^{\mathrm{err}} \\ P_3^{\mathrm{err}} \end{bmatrix} = \begin{bmatrix} |y_1^{\mathrm{err}}|^2 \\ |y_2^{\mathrm{err}}|^2 \\ |y_3^{\mathrm{err}}|^2 \end{bmatrix} \quad (3.9)$$

To minimize the impact of chip fabrication errors on system performance as much as possible, this research proposes a method that combines phase error compensation with power error compensation, with the specific physical logic illustrated in Fig. 3.18. The optical power output by the silicon-based integrated DONN chip after phase error compensation can be calculated using Eqs. 3.10 and 3.11. Further power error compensation is applied on the foundation of phase error compensation. At

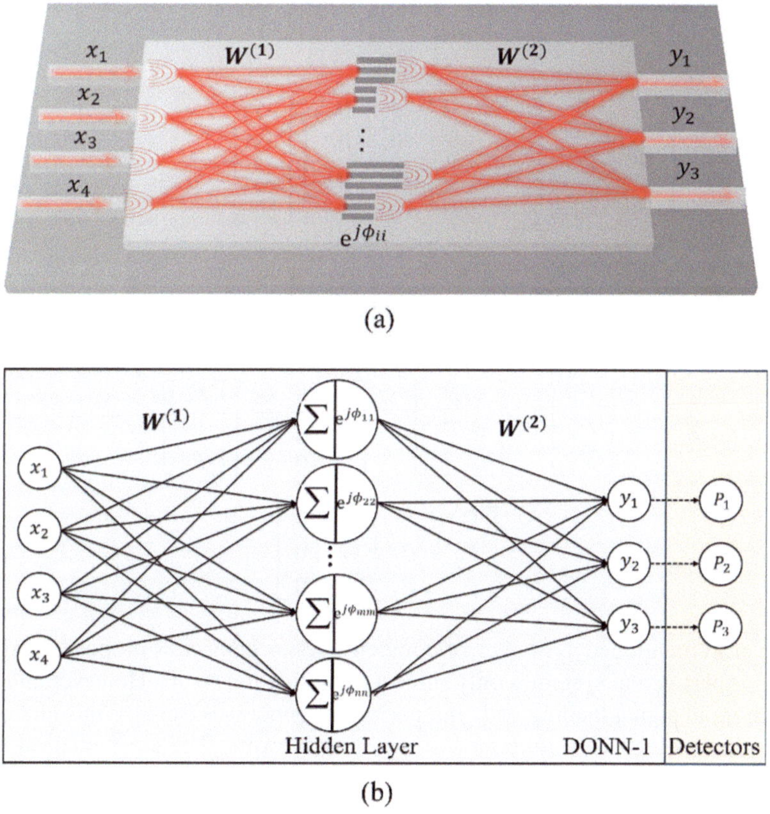

Fig. 3.17 Physical structure and mathematical logic diagram of silicon-based integrated DONN. **a** Schematic diagram of the silicon-based integrated DONN structure; **b** Logical diagram of the silicon-based integrated DONN

this point, the optical power output by the silicon-based integrated DONN chip can be calculated using Eq. 3.12. The purpose of performing phase error compensation and power error compensation is to make the optical power at each output port of the silicon-based integrated DONN chip as consistent as possible with the theoretical design, that is, to minimize the optical power error between the corresponding output ports, as shown in Eq. 3.13. Since the optical power at each output port of the silicon-based integrated DONN chip does not exhibit a fixed proportional relationship during simulation and experimental testing, this research adopts the final prediction result of the classification task, namely the prediction accuracy on the test set, as the criterion to judge the effectiveness of the system error compensation scheme during the system error compensation stage.

$$\begin{bmatrix} y_1' \\ y_2' \\ y_3' \end{bmatrix} = \begin{bmatrix} \mathbf{W}^{(2)} \end{bmatrix}_{3\times m}^{T} \begin{bmatrix} e^{j\phi_{11}\text{err}} & 0 & 0 & \cdots & 0 \\ 0 & e^{j\phi_{22}\text{err}} & 0 & \cdots & 0 \\ \vdots & \ddots & \ddots & \ddots & \vdots \\ 0 & \cdots & 0 & \ddots & 0 \\ 0 & \cdots & 0 & 0 & e^{j\phi_{mm}\text{err}} \end{bmatrix} \begin{bmatrix} \mathbf{W}^{(1)} \end{bmatrix}_{m\times 4}^{T} \begin{bmatrix} e^{j\varphi_{11}} & 0 & 0 & 0 \\ 0 & e^{j\varphi_{22}} & 0 & 0 \\ 0 & 0 & e^{j\varphi_{33}} & 0 \\ 0 & 0 & 0 & e^{j\varphi_{44}} \end{bmatrix} \begin{bmatrix} x_1 \\ x_2 \\ x_3 \\ x_4 \end{bmatrix} \tag{3.10}$$

$$\begin{bmatrix} P_1' \\ P_2' \\ P_3' \end{bmatrix} = \begin{bmatrix} |y_1'|^2 \\ |y_2'|^2 \\ |y_3'|^2 \end{bmatrix} \tag{3.11}$$

$$\begin{bmatrix} P_1'' \\ P_2'' \\ P_3'' \end{bmatrix} = \begin{bmatrix} \alpha_1 P_1' \\ \alpha_2 P_2' \\ \alpha_3 P_3' \end{bmatrix} = \begin{bmatrix} \alpha_1 & 0 & 0 \\ 0 & \alpha_2 & 0 \\ 0 & 0 & \alpha_3 \end{bmatrix} \begin{bmatrix} P_1' \\ P_2' \\ P_3' \end{bmatrix} \tag{3.12}$$

$$\begin{bmatrix} P_1'' - P_1 \\ P_2'' - P_2 \\ P_3'' - P_3 \end{bmatrix} \approx \begin{bmatrix} 0 \\ 0 \\ 0 \end{bmatrix} \tag{3.13}$$

The aforementioned system error compensation scheme was proposed assuming that the experimental system is only subject to fabrication errors. However, in reality, both the input signal loading and chip packaging processes can also introduce additional errors, which constitute part of the system errors. The errors caused by the signal loading and packaging processes can also be corrected to some extent by the proposed system error compensation scheme. Specifically, the errors generated during the input signal loading process can be corrected during the phase error compensation stage, while the errors introduced during the packaging process are essentially fixed offsets and can be compensated through the combined action

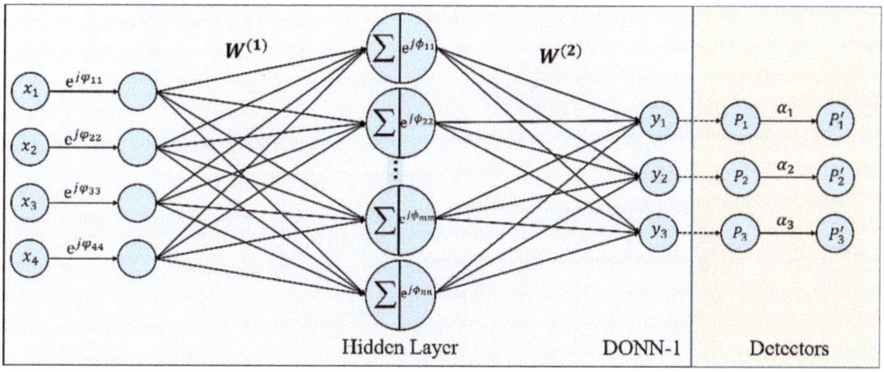

Fig. 3.18 Schematic diagram of system error compensation logic for the silicon-based integrated DONN chip

of phase error compensation and power error compensation. Thus, the proposed system error compensation scheme can effectively compensate for the system errors introduced during the testing process of the experimental system to a certain extent.

It is particularly noteworthy that the phase shifts corresponding to the input signal features are transformed into corresponding voltages according to the voltage-phase relationship illustrated in Fig. 3.16, thereby successfully accomplishing the loading process of input signal features onto the optical phase domain. The apparatus required for the system error compensation scheme is the same as the experimental testing apparatus for the silicon-based integrated DONN chip. Once the online system error compensation process is completed, a set of optimal voltage compensation values $\Delta V = (\Delta V_1, \Delta V_2, \Delta V_3, \Delta V_4)$ and a set of optimal power compensation coefficients $\boldsymbol{\alpha} = (\alpha_1, \alpha_2, \alpha_3)$ are obtained. Once this compensation process is completed, it exists as a fixed value in subsequent sample testing and does not introduce additional effects on the experimental testing system. Figure 3.19 illustrates the experimental system error compensation apparatus and procedure.

Due to fabrication errors in the silicon-based integrated DONN chips, the physical processes of light propagation within the chips cannot be accurately represented analytically. Therefore, during the error compensation stage, non-gradient descent algorithms and optimization search algorithms are employed to optimize the phase compensation factors ΔV_m and the power compensation coefficients α_m. The non-gradient descent algorithm used in this system error compensation process is the particle swarm optimization (PSO) algorithm, with the specific algorithm flow shown in Fig. 3.20. System error compensation can be regarded as an optimization problem, to enable the output power of the silicon-based integrated DONN chip to better perform the classification task of the IRIS Plants Dataset. Thus, the entire optimization problem can be formulated as the extremum problem of Eq. 3.14. The solution to the extremum problem of Eq. 3.14 will be carried out in two parts: phase error compensation and power error compensation.

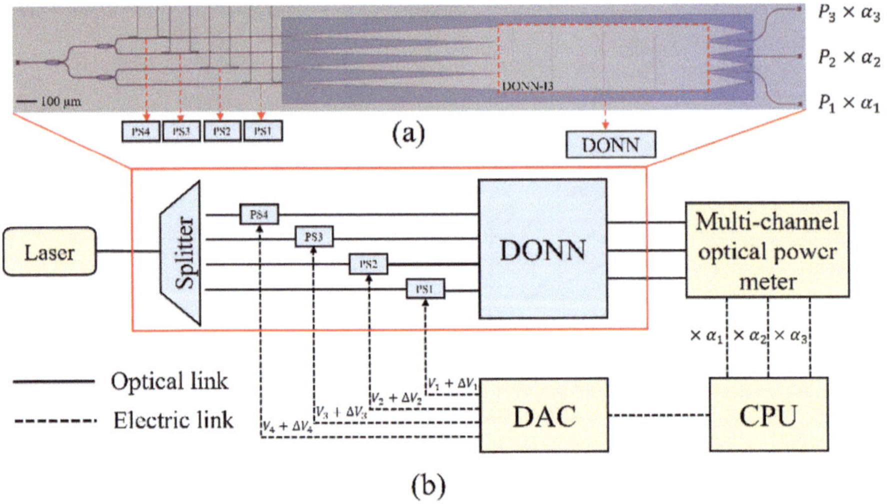

Fig. 3.19 Experimental system error compensation apparatus for the silicon-based integrated DONN chip [11]. Here, PS denotes the phase shifter; DAC represents the digital-to-analog converter; CPU is an abbreviation for the central processing unit. V_m and ΔV_m are voltage values, where ΔV_m is the phase compensation term; α_m is the power compensation coefficient, $m = 1, 2, 3, 4, \cdots$

$$\underset{\Delta \boldsymbol{V}^i, \boldsymbol{\alpha}^i}{\arg\min} \sum_{\dim(\boldsymbol{T})} \left\{ \left[\boldsymbol{Chip}\left(\boldsymbol{V}_{\text{signal}} + \Delta \boldsymbol{V}^i \right) \odot \boldsymbol{\alpha}^i \right] - \boldsymbol{T} \right\}^2 \tag{3.14}$$

where $\boldsymbol{V}_{\text{signal}}$ represents a set of voltage values corresponding to the features of the input signal; $\Delta \boldsymbol{V}^i$ represents a set of compensation voltage values that are continuously optimized during the phase error compensation process; $\odot$ denotes the element-wise multiplication; $\boldsymbol{\alpha}^i$ represents a set of power compensation coefficients that are continuously optimized during the power error compensation process. $\boldsymbol{T}$ represents a set of labels for the classification dataset, which takes the form of (1,0,0), (0,1,0), and (0,0,1). The goal of the optimization process is to obtain the voltage compensation values $\Delta \boldsymbol{V}^i = (\Delta V_1, \Delta V_2, \Delta V_3, \Delta V_4)$ and power compensation coefficients $\boldsymbol{\alpha}^i = (\alpha_1, \alpha_2, \alpha_3)$ that minimize Eq. 3.14.

PSO is a non-gradient descent algorithm [12–14] inspired by the foraging behavior of bird flocks, bee swarms, or ant colonies. This subsection will illustrate it using the example of a bird flock foraging. For instance, consider a group of woodpeckers in a forest whose objective is to find the location with the most insects. In a completely unfamiliar forest, the woodpeckers fly in randomly determined directions at random speeds to begin their search. During the foraging process, each woodpecker can remember where it found the most insects. At the same time, each woodpecker knows where the other woodpeckers in the flock have found the most insects. At this point, each woodpecker must decide whether to return to where it found the most insects or head toward where other woodpeckers in the flock have found the most insects. The conflicted woodpecker accelerates in both directions simultaneously,

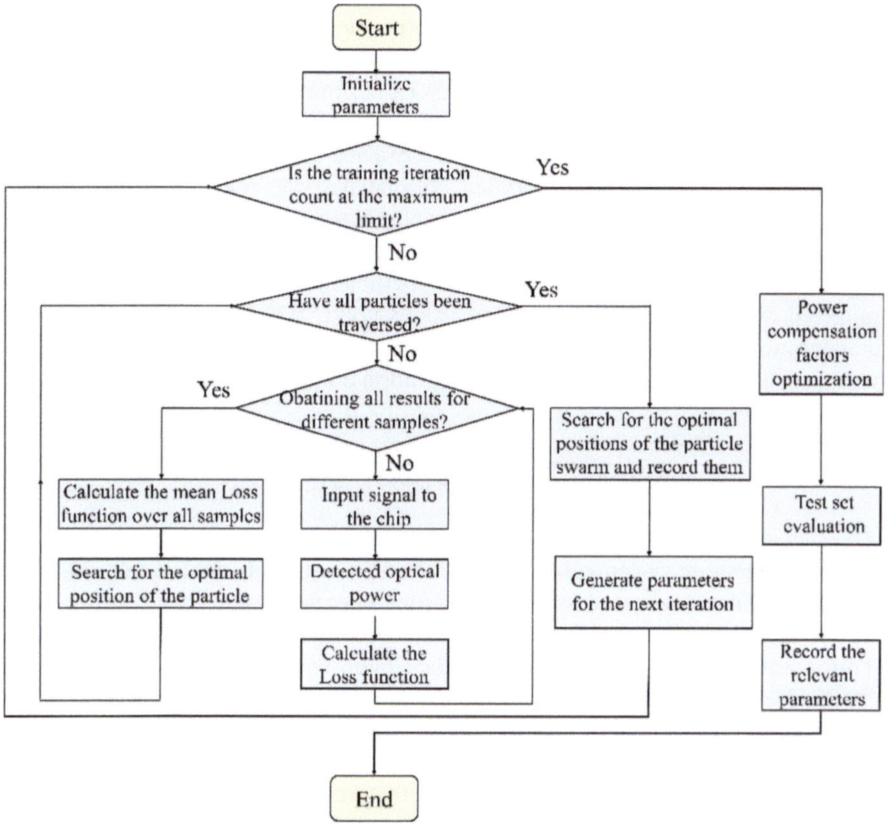

Fig. 3.20 Flowchart of experimental system error compensation for silicon-based integrated DONN chips based on particle swarm optimization algorithm

causing its flight trajectory to change due to the influence of these two factors. As it flies along with the new trajectory, the woodpecker may discover a location with an even higher concentration of insects. Consequently, it will continue to adjust its flight path based on the two new locations. Gradually, the entire flock will be drawn to the location with the most insects. In this process, if any woodpecker in the flock discovers a new insect hotspot that surpasses the previously identified location with the most insects, the flock will be attracted to this new hotspot. The woodpeckers will continuously revisit the locations where they previously found the most insects, hoping to find even more. Through this relentless search, most of the woodpeckers will eventually congregate in the forest with the highest concentration of insects.

As shown in Fig. 3.20, the system error compensation is divided into two stages. The first stage is the phase error compensation based on the PSO algorithm, which can find the optimal solution for phase compensation. On this basis, the second stage, power error compensation, is carried out to find the optimal power compensation coefficients. The collaboration of these two stages can yield the optimal compensation parameters for the entire system. During the phase compensation stage, the power compensation coefficients are fixed at $\boldsymbol{\alpha}^i = (1,1,1)$. The task is to optimize the phase compensation parameters $\Delta \boldsymbol{V}^i$ using the PSO algorithm. The optimization objective is to find a set of voltage compensation values $(\Delta V_1, \Delta V_2, \Delta V_3, \Delta V_4)$ that minimize the value of Eq. 3.15. Additionally, the performance of the optimal particle in each generation is recorded. These recorded optimal particles and the global optimal particle will participate in the power error compensation process in the second stage. In the power error compensation stage, an optimization search algorithm is employed to obtain the optimal power compensation coefficients $(\alpha_1, \alpha_2, \alpha_3)$. Ultimately, the optimized phase compensation voltage values $(\Delta V_1, \Delta V_2, \Delta V_3, \Delta V_4)$ and power compensation coefficients $(\alpha_1, \alpha_2, \alpha_3)$ will lead to the minimum value of Eq. 3.14, thereby completing the system error compensation for the entire experimental system setup.

$$\underset{\Delta \boldsymbol{V}^i}{\arg\min} \sum_{\dim(\boldsymbol{T})} \left\{ \boldsymbol{Chip}\left(\boldsymbol{V}_{\text{signal}} + \Delta \boldsymbol{V}^i\right) - \boldsymbol{T} \right\}^2 \tag{3.15}$$

During the entire system error compensation process, only the training set of the IRIS Plants Dataset was used for compensation. After the system error compensation was completed, the test set of the IRIS Plants Dataset was classified and predicted. At this point, the prediction accuracies of the silicon-based integrated DONN-I1 and DONN-I3 chips on the test set of the IRIS Plants Dataset were 86.7% and 90.0%, respectively. Figure 3.21 shows the test results of the silicon-based integrated DONN-I1 and DONN-I3 chips on the test set of the IRIS Plants Dataset before and after system error compensation. It is worth noting that, according to the experimental test results, the silicon-based integrated DONN-I1 and DONN-I3 chips performed well in predicting the class of Setosa both before and after system error compensation, which is consistent with the simulation validation (As shown in Fig. 3.8, during the simulation process, the prediction results of the silicon-based integrated DONN-I1 and DONN-I3 chips for Setosa were all correct. However, Versicolor and Virginica were more likely to be misclassified as other classes). Under the current structure parameters of the silicon-based integrated DONN, the experimental test results and simulation results show consistent performance in predicting the test set of the IRIS Plants Dataset.

In Fig. 3.21, red dots represent the test results of samples with input signal features belonging to Setosa, blue dots represent the test results of samples with input signal features belonging to Versicolor, and purple dots represent the test results of samples with input signal features belonging to Virginica. Additionally, the correct prediction results for sample numbers 1–10 should be Setosa, meaning that the output optical power value corresponding to the red dot should be the highest. For sample numbers 11–20, the correct prediction results should be Versicolor, meaning that the output optical power value corresponding to the blue dot should be the highest. For sample numbers 21–30, the correct prediction results should be Virginica, meaning that the output optical power value corresponding to the purple dot should be the highest. For example, in Fig. 3.21a, for sample numbered 5, the power value of the red dot

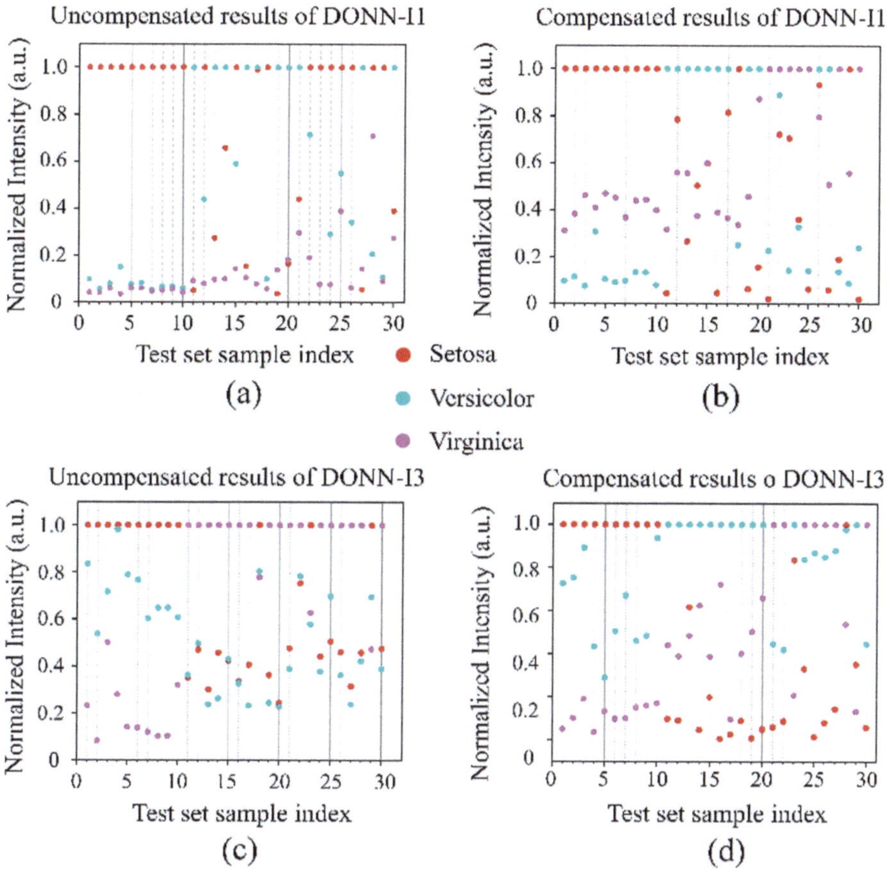

Fig. 3.21 Experimental test results of silicon-based integrated DONN-I1 and DONN-I3 chips

is the highest, so the prediction result is correct. However, for the sample numbered 15, the power value of the blue dot should be the highest, but the power value of the red dot is the highest at this time, so the prediction result is incorrect. Figure 3.21c, d show the experimental test results of the silicon-based integrated DONN-I3 chip, and the data analysis is the same as that in Fig. 3.21a, b. Additionally, the intensity of light is displayed after normalization for individual samples, and the normalization unit "a.u." is an abbreviation for an arbitrary unit.

Figure 3.22 shows the confusion matrices of the silicon-based integrated DONN-I1 and DONN-I3 chips for the IRIS Plants Dataset test set before and after system error compensation. Figure 3.23 illustrates the classification prediction results for Versicolor by the silicon-based integrated DONN-I3 chip before and after system error compensation.

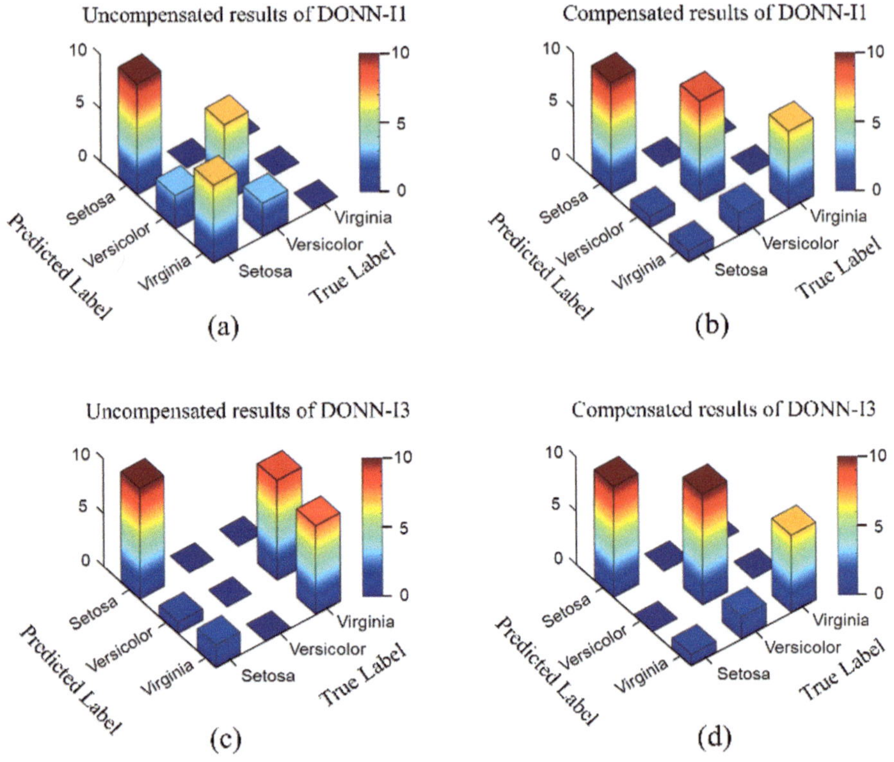

Fig. 3.22 Experimental test results of confusion matrices for silicon-based integrated DONN-I1 and DONN-I3 chips [11]. **a/b** Confusion matrices for the experimental test results of the integrated DONN-I1 chip before/after system error compensation. **c/d** Confusion matrices for the experimental test results of the silicon-based integrated DONN-I3 chip before/after system error compensation

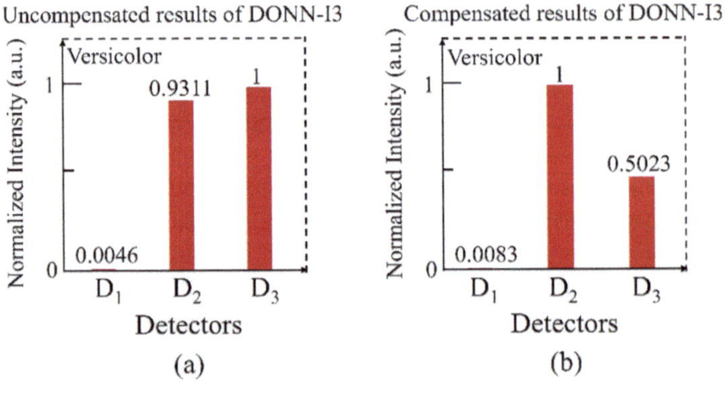

Fig. 3.23 Comparison of test results for the silicon-based integrated DONN-I3 chip before and after system error compensation [11]. **a** and **b** show the experimental prediction results for Versicolor by the integrated DONN-I3 chip before and after system error compensation, respectively

3.2.5 *Performance Analysis*

The preceding content has completed the fabrication and experimental testing of silicon-based integrated DONN chips. This subsection will analyze and discuss the performance of the silicon-based integrated DONN-I3 chip in terms of computing speed, computing delay, computing power consumption, and computing capacity. First, the computing speed of the silicon-based integrated DONN-I3 chip can be calculated using Eq. 3.1 and is found to be 1.38×10^{16} OPS. Regarding computing delays, Eq. 3.2 can be referenced and combined with Fig. 3.24 to calculate the delay of silicon-based integrated DONN-I3. Specifically, the maximum group refractive index, represented as $\max(n_{g1}, n_{g2}, n_{g3})$, is employed to provide an approximate estimation of the light propagation delay through the DONN-I3. Based on this approach, the maximum delay will not exceed 48.54 ps.

Regarding the calculation of energy consumption, the laser used in this study has an output power of 32 mW at 1.55 μm wavelength, and the power required for the thermo-optic phase shifter to achieve a 2π phase shift is about 30 mW. The input signal loading of the DONN-I3 chip integrated on a silicon substrate requires four phase shifters. The calculation gives the computing speed of the silicon-based integrated DONN-I3 chip to be about 1.38×10^{16} OPS. Therefore, the computing energy consumption of the silicon-integrated DONN-I3 chip is about 1.1×10^{-17} J/Operation. For the estimation of computing capacity, the theoretical calculation shows that the current silicon-based integrated DONN chip has a computing capacity of about 4.6×10^{16} OPS/mm^2.

This section compares the performance of the silicon-based integrated DONN-I3 chip with that of other silicon-integrated ONNs, as shown in Table 3.3. Here, CUs represent computing units. The silicon-based integrated DONN-I3 chip fabricated in this study has significant advantages in footprint, throughput, computing energy consumption, integration degree, and computing capacity compared with other silicon-based integrated ONN chips.

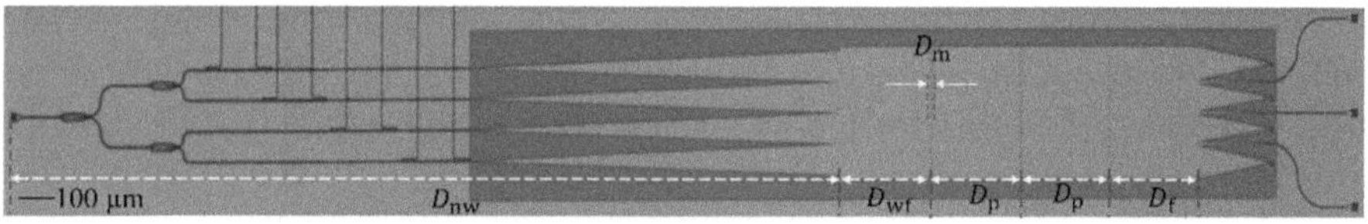

Fig. 3.24 Optical microscope image of the silicon-based integrated DONN-I3 chip

Table 3.3 Comparison of partial performance of DONN-I3 chip with other silicon-based integrated ONNs

Works	Footprint (mm^2)	Throughput (TOPS)	Energy consumption (J/Operation)	Integration degree (CUs/mm^2)	Theoretical computing capacity (OPS/mm^2)
Ref. [3]	0.68	6.4	7.66×10^{-14}	< 10	9.41×10^{12}
Ref. [15]	0.36	21.6	2.14×10^{-13}	< 10	6.00×10^{13}
Ref. [16]	1.20	32.0	1.41×10^{-15}	< 20	2.67×10^{13}
Ref. [17]	2.12	30.0	3.07×10^{-15}	< 25	1.42×10^{13}
Ref. [18]	6.07	28.8	5.90×10^{-15}	< 5	1.80×10^{12}
DONN-I3	0.30	1.38×10^{4}	1.10×10^{-17}	$\sim 2 \times 10^{3}$	4.60×10^{16}

3.3 Ten Classification Silicon-Based Integrated DONN Chip

In Sects. 3.1 and 3.2, silicon-based integrated DONN chips for binary and ternary classification tasks were designed, and their performance was validated through simulation and experimentation. This section will address a more complex classification task, the MNIST dataset, to further verify the performance of DONN chips designed based on the silicon-based integrated DONN model. The MNIST dataset [19] is a ten classification task dataset, which includes 10 different handwritten digits from 0 to 9. The training set consists of 60,000 images, and the test set contains 10,000 images, each with a size of 28 × 28 pixels. Before designing the silicon-based integrated DONN chip based on this dataset, preprocessing of the data is required. First, the 28 × 28-pixel image is converted into a 784 × 1 vector. Since 784 features are still too many for the silicon-based integrated DONN chip to handle simultaneously, this research further compresses the 784 × 1 vector into a 10 × 1 vector using a fully connected network. Thus, the input feature quantity for the silicon-based integrated DONN chip is reduced to 10. At this point, the MNIST training set is compressed into a 60,000 × 10 matrix, and the test set forms a 10,000 × 10 matrix. During the training process of the structure parameters of the silicon-based integrated DONN chip, the 10 features of the preprocessed MNIST dataset are loaded onto the phase of light, serving as the input signal for the silicon-based integrated DONN. The input signal is still realized through 10 thermo-optic phase shifters to load the phase features. For output, the handwritten digit category is determined by the intensity of light detected by 10 monitors D_i ($i = 0,1, \cdots, 9$).

3.3.1 Design of Structures

In this optimized design of the silicon-based integrated DONN chip, each hidden layer includes 70 SSSDGs (a total of 210 SSSDs), with a periodic arrangement of 1.5 μm, meaning the total length of the SSSDG array is 105 μm. The distance from the input interface of the silicon slab waveguide (i.e., the input interface of the silicon-based integrated DONN) to the first hidden layer (SSSDG array) is 250 μm. The spacing between adjacent hidden layers is also 250 μm, and the distance from the last hidden layer to the output interface of the silicon slab waveguide (i.e., the output interface of the silicon-based integrated DONN) is 250 μm as well. At the output interface of the silicon slab waveguide, 10 monitors (D_0 to D_9) are placed, each with a width of 8 μm and a center-to-center spacing of 8 μm between adjacent monitors. To ensure that the output signal of the silicon-based integrated DONN chip is consistent with the training process after fabrication, it is necessary to consider the mode selection process when the optical field is output. Therefore, mode selection for the light received in the monitor region must be considered during the training of the silicon-based integrated DONN chip. In this training process, the mode overlap integral is utilized to filter the TE_0 mode optical field, and the specific implementation can be referred to the content introduced in Sect. 3.2.1. When the hyperparameters of the silicon-based integrated DONN chip are determined, this section will optimize the phase weight parameters of the chip with different numbers of hidden layers. Table 3.4 provides the blind test accuracy of silicon-based integrated DONNs with different numbers of hidden layers on the MNIST test set.

In the table, DONN-M*m* indicates that the silicon-based integrated DONN chip designed for the MNIST dataset has *m* hidden layers. For example, DONN-M3 indicates that the silicon-based integrated DONN has three hidden layers. Table 3.4 shows that when the number of hidden layers exceeds three, the performance improvement of the silicon-based integrated DONN is relatively small. Therefore, this research selects the silicon-based integrated DONN-M3 for fabrication and experimental validation.

Figure 3.25 presents the phase weight parameters of the silicon-based integrated DONN-M3 chip. Since each hidden layer contains 70 SSSDGs, corresponding to 70 phase weight parameters, the phase weight parameters of each hidden layer are displayed as a 2 × 35 matrix of pixels in Fig. 3.25.

Table 3.4 Prediction results on the MNIST test set

DONN-M*m*	Prediction accuracy of the silicon-based integrated DONN model (%)
DONN-M1	72.6
DONN-M2	95.5
DONN-M3	96.3
DONN-M4	96.5
DONN-M5	96.1

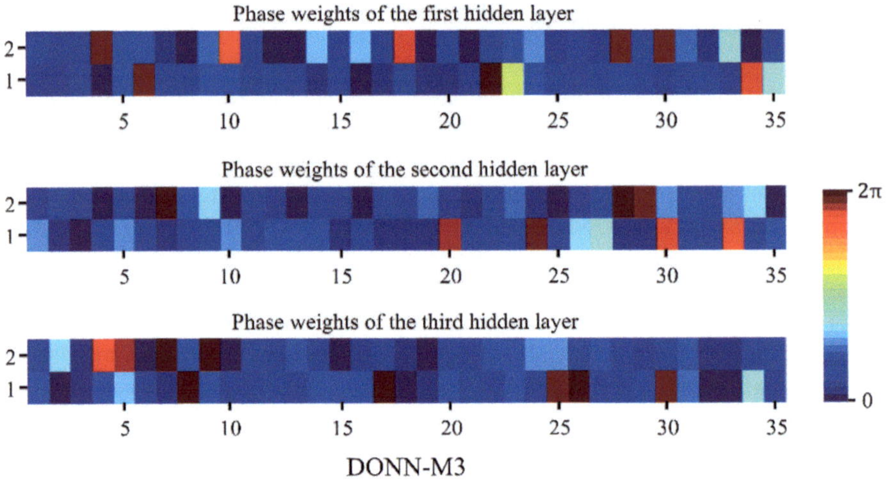

Fig. 3.25 Display of phase weights on each hidden layer of the silicon-based integrated DONN-M3 chip

Figure 3.26a illustrates the process of the loss function gradually converging with increasing optimization iterations during the training of the silicon-based integrated DONN-M3 chip. It presents the blind test results on the MNIST test set during this process. Figure 3.26b shows the confusion matrix for the classification prediction on the MNIST test set after the structure parameters of the silicon-based integrated DONN-M3 chip have been optimized.

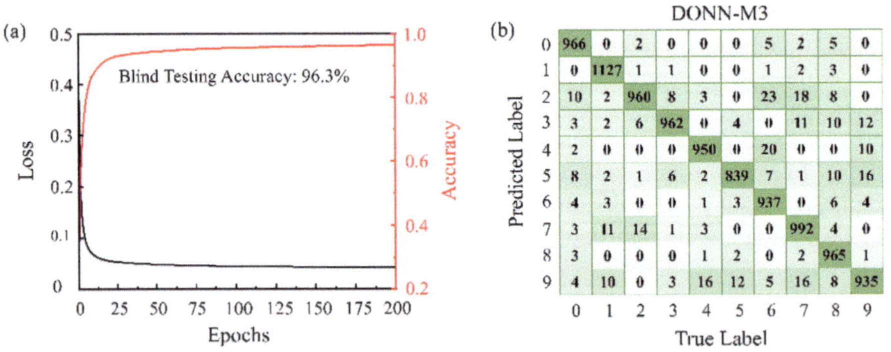

Fig. 3.26 Silicon-based integrated DONN-M3 chip training process and experimental test results demonstration. **a** Loss function convergence process and MNIST test set blind test results; **b** MNIST test set confusion matrix

3.3.2 Experimental Tests and Results Analysis

The fabrication process of the silicon-based integrated DONN-M3 chip in this study is essentially the same as that described in Sect. 3.2.3, with the difference being that electron-beam lithography is used for pattern transfer in this instance, offering higher processing accuracy than the previous electron-beam exposure technology. The minimum feature size of the silicon-based integrated DONN-M3 chip remains at 200 nm. According to the fabrication conditions of the chip foundry, the metal material used for the thermo-optic phase shifters in this study is titanium (Ti), while aluminum (Al) is used for the metal interconnects and electrodes. Finally, an 800 nm thick SiO_2 protective layer is deposited on the surface of the chip.

Figure 3.27 shows the optical microscope image of the silicon-based integrated DONN-M3 chip. The two devices in the figure have identical structure parameters, with one device serving as a backup. During the operation of the chip, light from the laser (with a wavelength of 1.55 μm) enters the silicon waveguide through the input grating coupler. Subsequently, by applying specific voltages to the metal electrodes, the connected thermo-optic phase shifters are heated. This heating, through its thermal effect, alters the effective refractive index of the silicon waveguide beneath the phase shifter, causing a specific phase shift in the light passing through that section of the waveguide. This process loads the input signal features onto the phase of light. When the optical field carrying the input signal features propagates through the silicon-based integrated DONN-M3 chip, it undergoes interference and diffraction due to the phase modulation by the on-chip SSSDG array (including three hidden layers). This results in a specific optical field distribution at the output interface of the silicon slab waveguide. The light in the monitor region is then coupled out through the output coupler and detected by a multi-channel optical power meter. Finally, the category of the handwritten digit from the MNIST dataset can be determined based on the power levels at the output ports, thus completing the classification prediction for the input sample of the MNIST dataset.

Figure 3.27 is an optical microscope image in which the SSSDG array appears blurry. Therefore, a scanning electron microscope (SEM) was used to observe the morphology of the SSSDG array, as shown in Fig. 3.28. The figure shows that the four corners of the micro-slot are not strict right angles but rounded, which differs from the structure used in the simulation. Thus, errors introduced during chip fabrication are objectively present. The testing method for the ten classification silicon-based integrated DONN-M3 chip is the same as that for the ternary silicon-based integrated DONN chip (as shown in Figs. 3.13c, 3.15, and 3.19). First, the bare die test must be completed to confirm that the optical paths on the chip are all functioning properly. Then, the bare die is wire-bonded and packaged, with the process illustrated in Fig. 3.29.

After completing the packaging of the silicon-based integrated DONN-M3 chip, the next step is to perform system error compensation. Before proceeding with the system error compensation, obtaining the voltage-phase curve of the thermo-optic phase shifters in this chip fabrication is necessary. The testing method involves

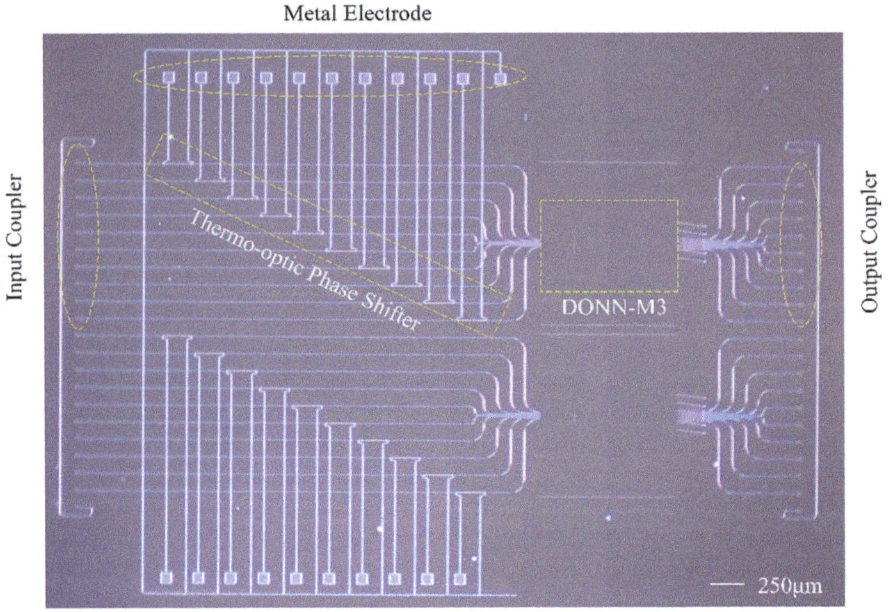

Fig. 3.27 Optical microscope image of the silicon-based integrated DONN-M3 chip

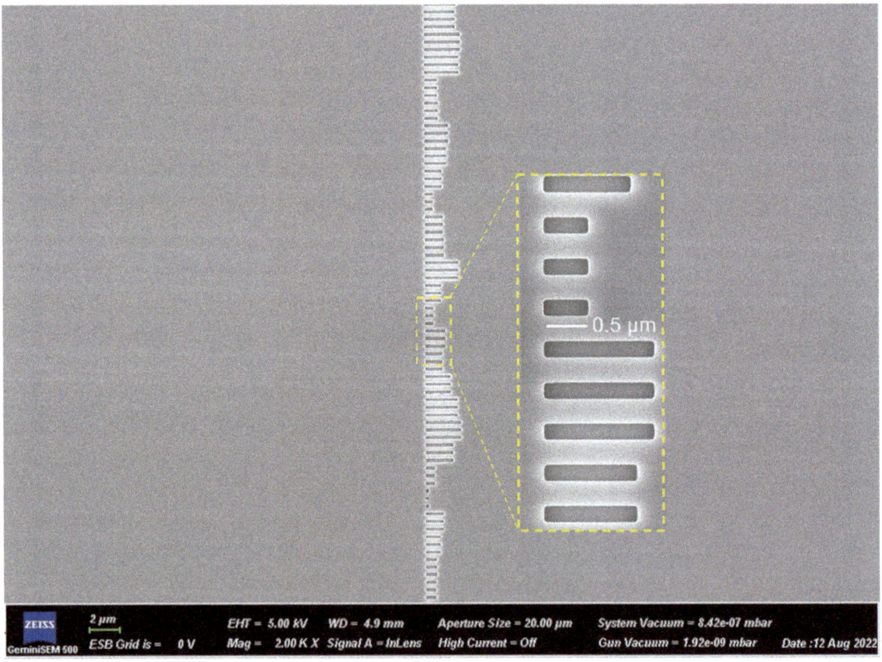

Fig. 3.28 SEM image of the silicon-based integrated DONN-M3 chip

Fig. 3.29 Silicon-based integrated DONN-M3 chip packaging process and results display. **a** DONN-M3 chip packaging process; **b** Close-up of wire bonding process; **c** Photo of silicon-based integrated DONN-M3 chip after packaging

sweeping the voltage on the phase shifter in one arm of an MZI to find the periodic change in output power with voltage, thereby calculating the voltage-phase curve shown in Fig. 3.30.

With the voltage-phase curve available, system error compensation can be carried out. Given the relatively high input dimensionality of the MNIST dataset, during the experimental process, 600 training samples were randomly selected from the 60,000 training samples, and 100 test samples were randomly selected from the 10,000 test samples. Even so, the online in-situ training compensation method still requires a significant amount of time. In addition, under the existing experimental conditions, since the optical path is affected by polarization, ambient temperature, and vibration during the signal loading stage, it is extremely difficult to ensure the experimental system's long-term stability. Therefore, the online in-situ training compensation process would be very challenging. Thus, for the system error compensation of the MNIST classification task, only the second stage of the compensation scheme mentioned in Sect. 3.2.4 was selected, namely, power error compensation. Since the MNIST classification task is more complex than the iris classification task and the system errors are more difficult to correct, a 10×10 fully connected layer network

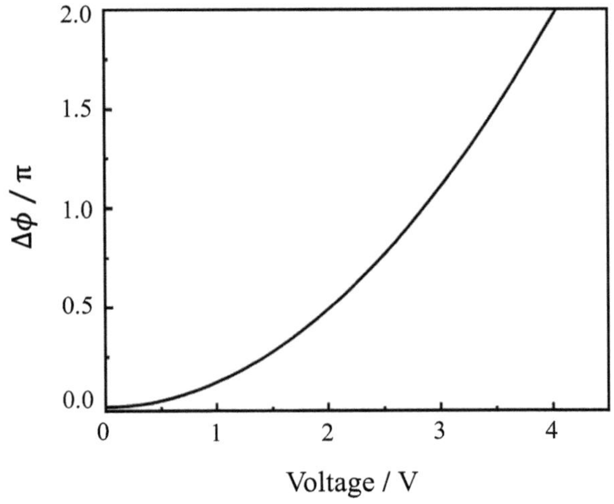

Fig. 3.30 Voltage-phase curve

W^{full} was set up after the output of the silicon-based integrated DONN-M3 chip to compensate for the system errors. The mathematical expression for this system error compensation process is shown in Eq. 3.16. The goal is to minimize the absolute difference between the compensated power $\boldsymbol{P}'$ and the ideal target power $\boldsymbol{P}$, i.e., $\left|\boldsymbol{P}' - \boldsymbol{P}\right| \approx 0$. The weights of this fully connected layer can be obtained through machine learning methods.

$$\begin{bmatrix} P_1' \\ P_2' \\ \vdots \\ P_{10}' \end{bmatrix} = \left[\boldsymbol{W}^{\text{full}}\right]^T_{10\times 10} \begin{bmatrix} P_1^{\text{err}} \\ P_2^{\text{err}} \\ \vdots \\ P_{10}^{\text{err}} \end{bmatrix} = \begin{bmatrix} w_{11} & \cdots & w_{1,10} \\ \vdots & \ddots & \vdots \\ w_{10,1} & \cdots & w_{10,10} \end{bmatrix}^T_{10\times 10} \begin{bmatrix} P_1^{\text{err}} \\ P_2^{\text{err}} \\ \vdots \\ P_{10}^{\text{err}} \end{bmatrix} \tag{3.16}$$

When the fully connected network is trained, the fully connected layer becomes a fixed network during the MNIST test set classification process. It is worth noting that only the randomly selected 600 MNIST training samples participated in the entire system error compensation process. After optimizing the system error compensation network, the blind test accuracy is 86.0% on the randomly selected 100 MNIST test set samples. Figure 3.31 shows the confusion matrix for the MNIST test set samples, and Fig. 3.32 presents the test results of the test set samples, where the normalized light intensity unit "a.u." is an abbreviation for an arbitrary unit.

3.3.3 Performance Analysis

The computing speed of the silicon-based integrated DONN-M3 chip is calculated based on Eq. 3.1. Assuming the detector bandwidth is 40 GHz, the computing speed

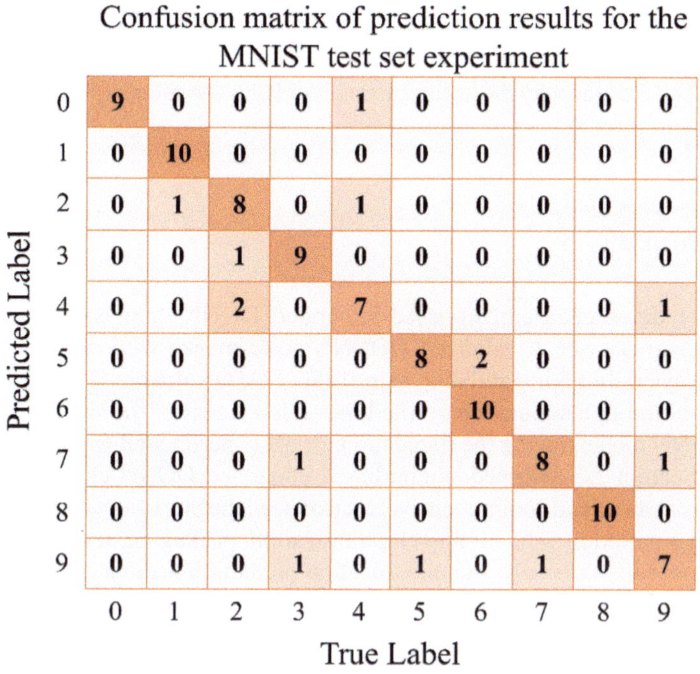

Fig. 3.31 Experimental prediction results of the silicon-based integrated DONN-M3 chip on the MNIST test set [11]

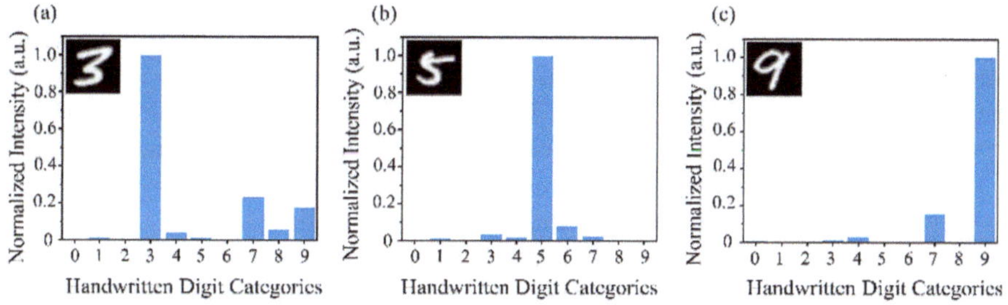

Fig. 3.32 Test results of silicon-based integrated DONN-M3 chip system with compensation [11]

of the silicon-based integrated DONN-M3 chip is approximately 7.84×10^{14} OPS. On this basis, considering that the input power of the laser is 32 mW, and there are 10 thermo-optic phase shifters in this chip, with each phase shifter requiring about 26 mW to achieve a 2π phase shift, the computing energy consumption of the chip is calculated to be approximately 3.72×10^{-16} J/Operation.

The difference between fabricating the silicon-based integrated DONN-I1 and DONN-I3 chips in Sect. 3.2 is that the beam splitting process for the silicon-based integrated DONN-M3 chip is completed off-chip. Therefore, the system testing process requires higher ambient temperature and stability conditions. Additionally, the polarization of each optical fiber can affect the system's performance, which is one of the key reasons why the accuracy rate is only 86.0% after experimental system error compensation.

3.4 Summary

Section 3.1 designed a silicon-based integrated DONN chip based on the UCI Heart Disease Dataset. It validated DONN's performance with different numbers of hidden layers through simulation using 2.5D FDTD software. This section theoretically and through simulation verified the effectiveness of the silicon-based integrated DONN model, laying the foundation for the experimental work in subsequent chapters. Section 3.2 designed silicon-based integrated DONN chips based on the IRIS Plants Dataset and comprehensively introduced the preparation from theoretical design and simulation validation to chip fabrication, packaging, testing, and system error compensation. Finally, the classification prediction accuracies of the silicon-based integrated DONN-I1 and DONN-I3 chips on the IRIS Plants Dataset test set reached 86.7% and 90.0%, respectively. It is worth noting that the test results after system error compensation are consistent with those in the simulation validation process. This conclusion demonstrates the feasibility and effectiveness of the system error compensation scheme. Combining this compensation scheme will significantly enhance the robustness of the silicon-based integrated DONN chips. In addition, the highest classification accuracy for the IRIS Plants Dataset in the simulation calculation was 90.0%, which is not as high as the classification accuracy of electrical neural networks. The reason may be due to the hyperparameters of the DONN structure, as the hyperparameters were not updated during the training process of the IRIS Plants Dataset classification network. Therefore, the fixed hyperparameters may affect the performance of the entire optical network. This issue can be resolved by optimizing the hyperparameters simultaneously during the training process.

To further validate the performance of the silicon-based integrated DONN model and the capability of silicon-based integrated DONN chips in handling more complex tasks, Sect. 3.3 designed a silicon-based integrated DONN chip based on the ten classification MNIST dataset and experimentally verified the performance of the silicon-based integrated DONN-M3 chip. Thus, the silicon-based integrated DONN chip has completed machine learning classification inference tasks of varying difficulty and scale.

Regarding the simulation results presented in Fig. 3.2f, which shows that the blind test accuracy on the test set does not significantly improve when the number of hidden layers in the silicon-based integrated DONN exceeds three, the reason for this phenomenon may be analyzed as follows: The silicon-based integrated DONN system is a linear system with limited inference capabilities. Therefore, when the prediction accuracy reaches 86.9%, it represents the optimal performance of the silicon-based integrated DONN. Even with the addition of more hidden layers, it is difficult to significantly improve the network's performance under the determined hyperparameters.

In addition, during the Iris classification prediction experiment, the optical power received by the three receiving ports of the multi-channel optical power meter mainly ranges from −50 dBm to −25 dBm. The insertion loss introduced by the packaged input and output couplers is approximately 11 dB. Under favorable conditions, the optical power at the target output port of the multi-channel detector will be relatively high (around −25 dBm) after applying voltages. Although the detected optical power loss is significant from the experimental results, the optical power difference between each port is generally more than 3 dBm, which does not affect the classification results. The root cause of the overall significant loss is that the light at the output interface is not fully utilized. For example, the output interface width of the silicon-based integrated DONN-I1 and DONN-I3 chips is 300 μm, while the width of each monitor is only 8 μm, totaling 24 μm for the three monitors. In this case, the light in other areas is not utilized. The same issue in the handwritten digit classification prediction experiment using the MNIST dataset also exists. In fact, the monitors on the output interface could cover the entire interface. However, the problem is that as the coverage area of the monitors increases, the number of SSSDGs (phase modulation units) in the SSSDG array is limited. In this case, the performance of the silicon-based integrated DONN will deteriorate because the limited number of modulation units also means limited control over the optical field. Therefore, if it is necessary to further expand the coverage area of the monitors while maintaining good performance of the silicon-based integrated DONN, it is essential to increase either the number of layers in the SSSDG array or the number of modulation units in each layer of the SSSDG array. Generally, when designing the parameters of a silicon-based integrated DONN, the complexity of the task to be solved is considered. If satisfactory classification results can be obtained with a smaller coverage area of the monitors, there is no need to further expand the coverage area of the monitors, nor is it necessary to add more SSSDG arrays or modulation units in the array to increase the optical power received by the monitors. As for the distribution of other optical fields (stray light) around the monitors, this stray light will not interfere with or alter the optical power received by the monitors. This is because the stray light around the monitors will enter the attenuators and dissipate, thus not affecting the performance of the silicon-based integrated DONN chip.

Overall, it is reasonable to design the most suitable silicon-based integrated DONN chip according to the specific requirements of different tasks. Moreover, the ability to effectively address machine learning inference tasks under limited conditions is also an aspect that needs to be focused on during the design process of silicon-based integrated DONN chips.

References

1. Detrano R, Janosi A, Steinbrunn W et al (1989) International application of a new probability algorithm for the diagnosis of coronary artery disease. Am J Cardiol 64(5):304–310
2. Fu T, Zang Y, Huang H et al (2021) On-chip photonic diffractive optical neural network based on a spatial domain electromagnetic propagation model. Opt Express 29(20):31924–31940
3. Shen Y, Harris N, Skirlo S et al (2017) Deep learning with coherent nanophotonic circuits. Nat Photonics 11(7):441–446
4. Ismaeel S, Miri A, Chourishi D (2015) Using the Extreme learning machine (ELM) technique for heart disease diagnosis. In: 2015 IEEE Canada international humanitarian technology conference (IHTC2015). IEEE, pp 1–3
5. Saxena K, Sharma R (2016) Efficient heart disease prediction system. Proc Comput Sci 85:962–969
6. Shah S, Batool S, Khan I et al (2017) Feature extraction through parallel probabilistic principal component analysis for heart disease diagnosis. Physica A 482:796–807
7. Vijayashree J, Sultana H (2018) A machine learning framework for feature selection in heart disease classification using improved particle swarm optimization with support vector machine classifier. Program Comput Softw 44:388–397
8. Haq A, Li J, Memon M et al (2018) A hybrid intelligent system framework for the prediction of heart disease using machine learning algorithms. Mob Inf Syst 2018(1):3860146
9. Saqlain S, Sher M, Shah F et al (2019) Fisher score and Matthews correlation coefficient-based feature subset selection for heart disease diagnosis using support vector machines. Knowl Inf Syst 58:139–167
10. Blake C (1998) UCI repository of machine learning databases. http://www.ics.uci.edu/~mlearn/MLRepository.html.
11. Fu T, Zang Y, Huang Y et al (2023) Photonic machine learning with on-chip diffractive optics. Nat Commun 14(1):70
12. Kennedy J, Eberhart R (1995) Particle swarm optimization. In: Proceedings of ICNN'95-international conference on neural networks. IEEE, vol 4, pp. 1942–1948
13. Eberhart R, Shi Y (1998) Evolving artificial neural networks. In: Proceedings of the international conference on neural networks and brain. PRC, vol 1, pp PL5-PLI3
14. Robinson J, Rahmat-Samii Y (2004) Particle swarm optimization in electromagnetics. IEEE Trans Antennas Propag 52(2):397–407
15. Zhang H, Gu M, Jiang X et al (2021) An optical neural chip for implementing complex-valued neural network. Nat Commun 12(1):457
16. Zhu H, Zou J, Zhang H et al (2022) Space-efficient optical computing with an integrated chip diffractive neural network. Nat Commun 13(1):1044
17. Zhao X, Lv H, Chen C et al (2021) On-chip reconfigurable optical neural networks
18. Feldmann J, Youngblood N, Karpov M et al (2021) Parallel convolutional processing using an integrated photonic tensor core. Nature 589(7840):52–58
19. LeCun Y (1998) The MNIST database of handwritten digits. http://yann.lecun.com/exdb/mnist/

Chapter 4
Research on Multifunctional Silicon-Based Integrated Diffractive Optical Neural Network

Abstract The silicon-based integrated DONN chips introduced in Chap. 3 are all designed for single tasks. Although they have advantages such as low latency and low power consumption in specific application scenarios, they still have limitations. First, the input signal loading is implemented using thermo-optic phase shifters, which means they cannot handle tasks with high-speed signal inputs. Second, they cannot deal with multiple tasks, limiting the applications of silicon-based integrated DONN chips. Third, they cannot directly handle tasks with serial input signals. Therefore, the main research content of this chapter is to design a multifunctional silicon-based integrated DONN chip. The specific arrangements are as follows: Sect. 4.1 will introduce the architecture of the multifunctional silicon-based integrated DONN network; Sect. 4.2 will introduce the serial-to-parallel conversion technology based on arrayed waveguide structures, which is a key method for realizing high-speed serial signal input; Sect. 4.3 will introduce the silicon-based integrated DONN designed based on multi-task learning and transfer learning methods, and verify the performance of a single silicon-based integrated DONN in solving multiple ten-classification tasks through numerical calculations. Finally, Sect. 4.4 will summarize the content of this chapter.

4.1 Network Architecture

Multifunctional silicon-based integrated DONNs mainly consist of the serial-to-parallel conversion part based on arrayed waveguide structures and the multi-task silicon-based integrated DONN part. The serial-to-parallel conversion part primarily serves to achieve high-speed serial signal input. On the other hand, the multi-task silicon-based integrated DONN part mainly performs inference computations for multiple tasks and provides the inference results. The network architecture of the multifunctional silicon-based integrated DONN is depicted in Fig. 4.1. A total of m signals is input in temporal sequence, loaded onto the amplitude of the input light with intervals of Δt (seconds), and then split into various waveguides by an optical beam splitter. After passing through the delay lines designed in each waveguide,

© Tsinghua University Press 2026
T. Fu, *Research on Silicon-Based Integrated Diffractive Optical Neural Network*, Springer Theses, https://doi.org/10.1007/978-981-95-3787-7_4

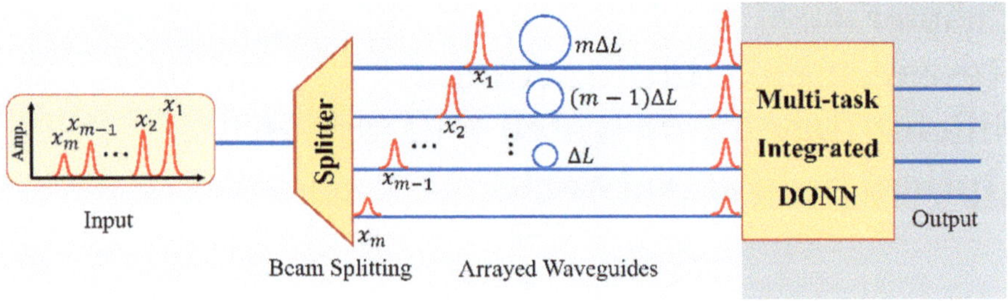

Fig. 4.1 Multifunctional silicon-based integrated DONN network architecture

the signals arrive simultaneously at the input interface of the multi-task silicon-based integrated DONN at a specific moment. Subsequently, the silicon-based integrated DONN performs inference computations on the input signals passively at that moment and provides the inference results through the varying optical intensities in the output waveguides. The feasibility of the multifunctional silicon-based integrated DONN network architecture will be verified separately by the serial-to-parallel conversion part and the multi-task silicon-based integrated DONN part in the following sections.

4.2 Serial-to-Parallel Conversion Technology Based on Arrayed Waveguide Structure

Looking back at the silicon-based integrated DONN chips implemented in Chaps. 2 and 3, it can be observed that the signal input is realized through the parallel loading of phase signals via thermo-optic phase shifters. The parallel signal loading through thermo-optic phase shifters is limited by the inherent response speed of the thermo-optic effect, which prevents the realization of high-speed signal modulation. In addition, for serial input signals, the data must first be parallelized on a computer. Then, the input signal features are loaded onto the phase of light in a parallel manner through the thermo-optic phase shifters as the input signal of the silicon-based integrated DONN. Moreover, as the number of parallel signal features increases, more thermo-optic phase shifters are required, inevitably leading to a rapid increase in the power consumption of the entire silicon-based integrated DONN system. In contrast, the serial-to-parallel conversion technology based on an arrayed waveguide structure is a technique that trades time for space. The signal loading achieved through this method does not rely on on-chip thermo-optic phase shifters but instead uses the delay lines of on-chip arrayed waveguides to realize parallel signal input. This meets the requirements for high-speed signal input and can also process serial input signals. The delay lines' length depends on the input signal's modulation rate; the higher the modulation rate, the shorter the delay lines designed in the arrayed waveguides.

Therefore, as the modulation rate continues to increase, the scale of serial-to-parallel conversion realized through arrayed waveguides will become more prominent.

4.2.1 Structure Design of Arrayed Waveguides

The key parameter in the structure design of the arrayed waveguide is the length of the delay line, which is related to the modulation rate of the input signal and the group refractive index of the waveguide. The length of the delay line can be calculated using Eq. 4.1.

$$\Delta L = \frac{c}{B \cdot n_g} \tag{4.1}$$

Here, ΔL represents the length difference between two adjacent arrayed waveguides; B is the modulation rate of the input signal; n_g is the group refractive index of the arrayed waveguides, which is used to calculate the time delay generated when ultrashort pulses propagate through the arrayed waveguides; c is the speed of light in vacuum.

This study used FDTD optical software to simulate the crosstalk between adjacent arrayed waveguides in the waveguide delay line to avoid crosstalk when the delay line bends. When the waveguide spacing is greater than 2 μm, the crosstalk between waveguides can be ignored. In this design, the center-to-center spacing between adjacent waveguides in the delay line bending region is set to 6 μm, so it can be determined that crosstalk will not occur between adjacent waveguides in the delay line bending region.

Figure 4.2 shows the calculation results of the bending loss of waveguides as a function of radius obtained using the MODE Solutions optical software. The thickness of the bending waveguide is 220 nm, and the width is 450 nm. The simulation results indicate that when the radius of the bending waveguide is larger than 4.5 μm, the bending loss approaches a lower value. In this research, the minimum bending radius of the arrayed waveguide delay lines is designed to be 10 μm. Therefore, the loss of the bending waveguide delay lines can be neglected. Given that the length design of the arrayed waveguide delay lines is intrinsically linked to the modulation rate of the input signal, this work presents the design of arrayed waveguide delay lines tailored explicitly for signal modulators operating at modulation rates spanning from 10 to 100 Gbps. The length of the arrayed waveguide delay lines was calculated for different modulation rates according to Eq. 4.1, which is shown in Table 4.1, with the value of n_g being 4.2.

From the results in Table 4.1, the higher the modulation rate, the shorter the delay line of the arrayed waveguide will be. Consequently, the loss caused by the waveguide delay will also be lower, and the scale of the serial-to-parallel conversion will be larger. At the current stage, assuming a modulation rate of 100 Gbps and a standard process loss of the arrayed waveguide of 1.5 dB/cm, the length of the

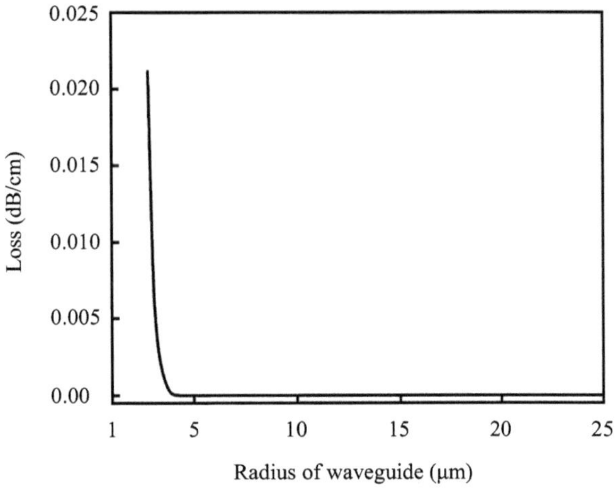

Fig. 4.2 Bending loss of waveguides as a function of different radii

Table 4.1 Arrayed waveguide delay lines for different modulation rates

Modulation rate of input signals (Gbps)	Shortest delay line of the arrayed waveguide (mm)
10	7.143
20	3.571
30	2.381
40	1.786
50	1.429
60	1.191
70	1.020
80	0.893
90	0.794
100	0.714

waveguide delay line is 0.7143 mm. When the scale of serial-to-parallel conversion reaches 1 × 100 channels, the total loss, including the beam splitting part (star coupler) and the arrayed waveguide delay line part, is approximately 31 dB. As the performance of modulators continues to improve, if the modulation rate increases to 1 Tbps, the length of the waveguide delay line will be 7.143×10^{-2} mm. If the standard process loss for fabricating the arrayed waveguide is reduced to 1 dB/cm at this time, when the serial-to-parallel conversion scale reaches 1 × 100 channels, the total loss, including the beam splitting part and the arrayed waveguide delay line part, will be approximately 21 dB. With the increase in the modulation rate of the input signal and the further reduction in the waveguide loss of semiconductor processes, the scale of serial-to-parallel conversion channels will continue to grow.

In this experiment, the modulation rate of the input signal is selected to be 10 Gbps. Therefore, the shortest delay line of the arrayed waveguide is 7.143 mm. The

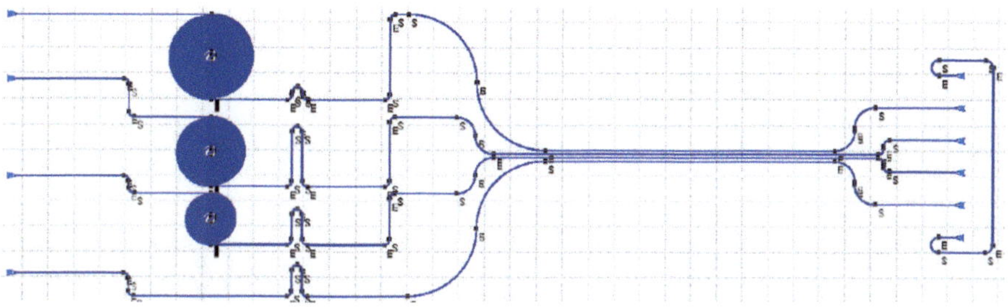

Fig. 4.3 Fabrication layout of the arrayed waveguides

lengths of the on-chip arrayed waveguide delay lines are 7.143 mm, 14.286 mm, and 21.429 mm, respectively. Figure 4.3 shows the process layout of the arrayed waveguide for implementing the serial-to-parallel conversion function. The serial-to-parallel conversion implemented in this experiment is 1 × 4 channels. The 1 × 4-star coupler (optical splitter) is located off-chip. This star coupler will work with the on-chip arrayed waveguide delay lines to achieve the serial-to-parallel conversion function.

4.2.2 Experimental Testing of the Arrayed Waveguide Structure

In this study, the fabrication of the arrayed waveguide was accomplished through electron beam lithography. The fabricated arrayed waveguide structure is shown in Fig. 4.4. To facilitate subsequent testing, the chip was packaged after the bare die performance test. The packaged device is shown in Fig. 4.5.

After the arrayed waveguide chip is packaged, the serial-to-parallel conversion experiment will be conducted using the experimental setup shown in Fig. 4.6. In this experiment, four temporal Gaussian pulses with different amplitudes generated by an arbitrary waveform generator (AWG) are loaded onto a continuous-wave laser using an amplitude modulator (AM). The modulation rate of the AM determines the time slots of the input signal pulses. In this study, the bandwidth of the AM is 10 Gbps, so the time slots of the input signal pulses are 100 ps. Subsequently, the light output from the laser is evenly split into four channels through a 1 × 4 fiber star coupler outside the chip and then coupled into the on-chip arrayed waveguides via four vertical grating couplers. Finally, the serial input signals in temporal order will arrive at the output interface in parallel and simultaneously at a certain moment after passing through the on-chip arrayed waveguides and will be captured by the detector.

As shown in Fig. 4.6, the arrayed waveguides are numbered from bottom to top as 1, 2, 3, and 4. During the experiment, the measured losses of the arrayed waveguides numbered 1 to 4 were 15.38 dB, 17.93 dB, 22.03 dB, and 25.2 dB, respectively. To ensure that the parallel signals have the same loss when they reach the output

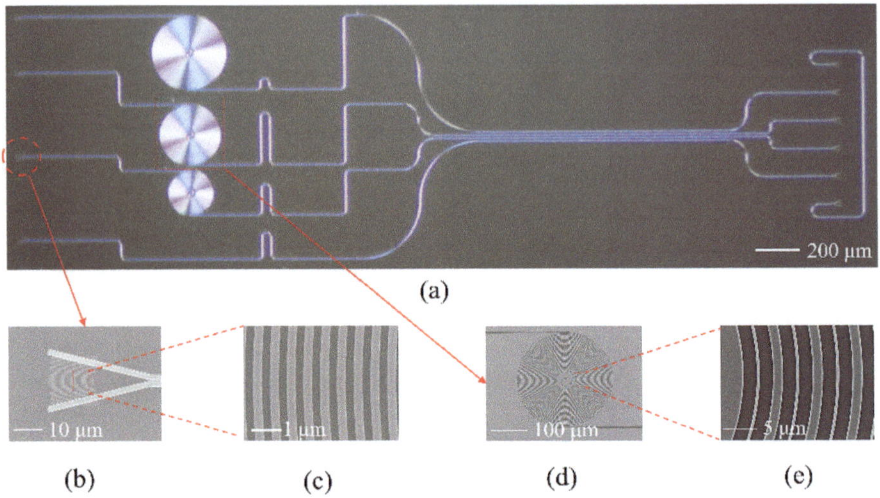

Fig. 4.4 Optical microscope image and partial SEM images during the fabrication of the arrayed waveguides [1]

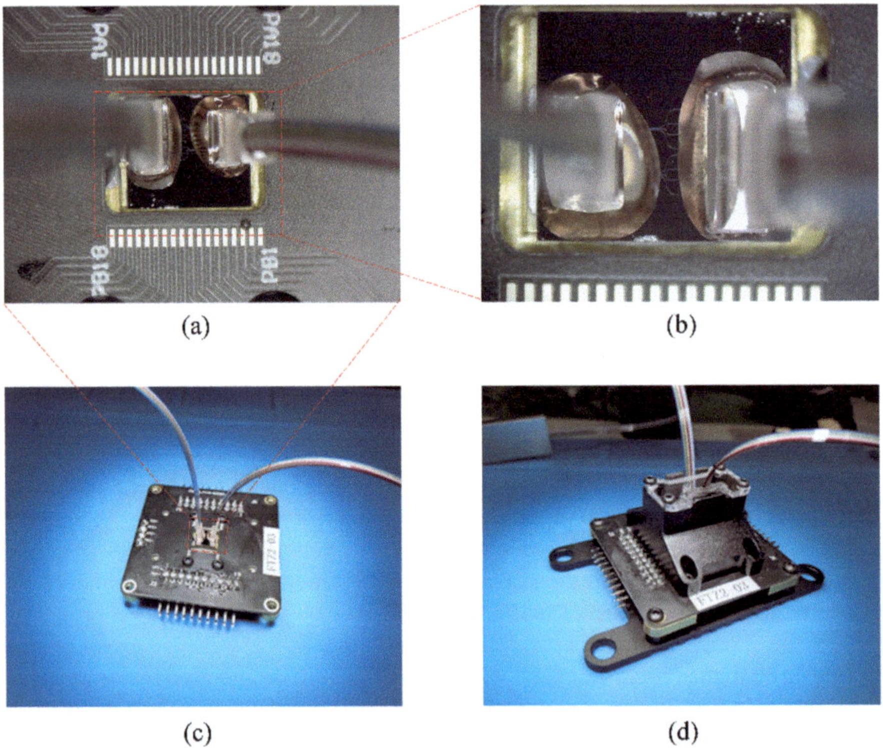

Fig. 4.5 The packaging process and results of the arrayed waveguide chip. **a**–**c** Illustrations of the packaging process; **d** Photograph of the packaged chip

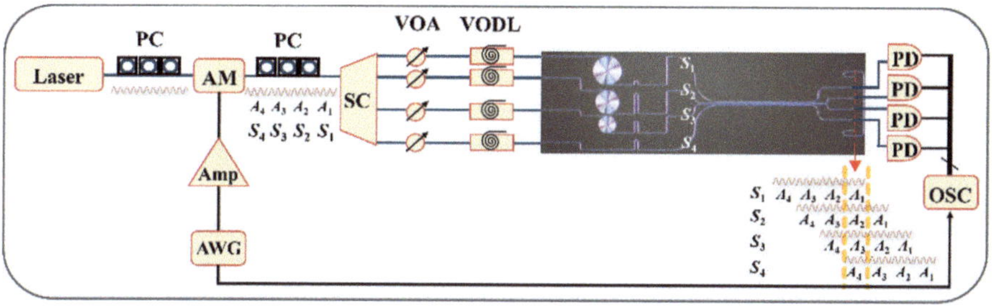

Fig. 4.6 Schematic diagram of the experimental setup and procedure for testing the packaged arrayed waveguide chip. Here, AWG denotes the arbitrary waveform generator, Amp represents the electric amplifier, AM is the amplitude modulator, SC is the 1 × 4 fiber star coupler, PD is the photodetector, OSC is the oscilloscope, A_i ($i = 1, 2, 3, 4$) represents the input signal, and S_i ($i = 1, 2, 3, 4$) marks the parallel output position of the serial input signals

interface, the experiments began by compensating for the fabrication errors in the delay lines. Variable optical delay lines were used to compensate the delay time between adjacent arrayed waveguides to 100 ps. On this basis, the losses of each arrayed waveguide delay line were further compensated using variable optical attenuators. According to the above test results, the loss introduced during the packaging process was approximately 6.5 dB per port. Therefore, the losses caused by the fabrication and packaging processes were relatively high. These losses and the delay line fabrication errors can be reduced through multiple iterations of chip fabrication.

Inevitable errors exist during the chip fabrication process. After compensating for the fabrication errors of the arrayed waveguides, the next step is to experimentally verify the performance of the arrayed waveguides in achieving serial-to-parallel conversion. When the serial input signals (A_1, A_2, A_3, A_4) are emitted from the AWG, modulated in intensity with a time slot interval of 100 ps by the AM, and then input into the 1 × 4 fiber star coupler, the serial signals are converted into parallel signals through the arrayed waveguide delay lines. The experimental results are shown in Fig. 4.7, where Fig. 4.7a shows the serial input signals, and Fig. 4.7b shows the optical intensity detection results of the serial input signals arriving at the output interface simultaneously after passing through the arrayed waveguide delay lines, i.e., the waveform displayed at 626.15 ps in Fig. 4.7b. This result demonstrates that the designed arrayed waveguides can effectively achieve the serial-to-parallel conversion function.

4.2.3 *Silicon-Based Integrated Diffractive Optical Neural Network with Serial-to-Parallel Conversion*

In Sect. 4.2.2, the experimental verification of the serial-to-parallel conversion performance of the arrayed waveguides was completed. Based on this experimental

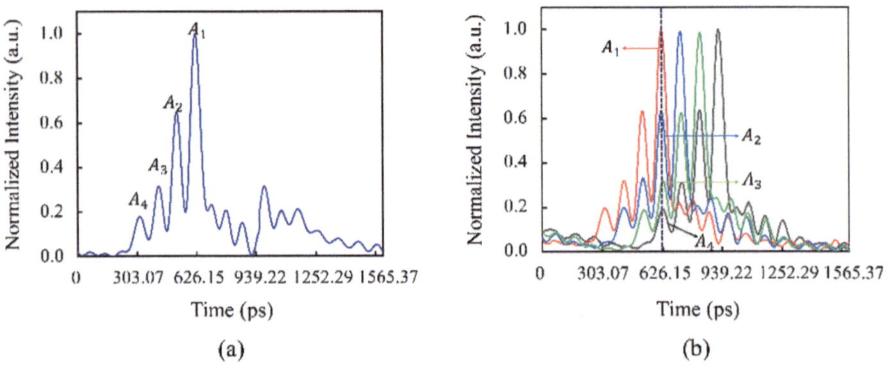

Fig. 4.7 The experimental results of the arrayed waveguide chip [1]. **a** Serial input signal; **b** Parallel output results at a certain moment after the serial input signal passes through the arrayed waveguide delay lines

premise, we will design a silicon-based integrated DONN for a vowel recognition task in this subsection. The performance verification of this silicon-based integrated DONN will be completed through numerical calculations. Figure 4.8 is a schematic diagram of the system (the on-chip DONN in the figure is a schematic diagram, not an experimental one). The system inputs the serial signal into the silicon-based integrated DONN in parallel after serial-to-parallel conversion using the arrayed waveguides.

The vowel recognition dataset [2] consists of 990 samples, which were obtained by having 15 individuals pronounce each of the 11 syllables six times, resulting in 90 recordings for each syllable. The first 48 recordings of each syllable were allocated to the training set, while the remaining 42 recordings were assigned to the testing set. Each syllable is characterized by 10 extracted features. In summary, the dataset contains 528 training samples and 462 testing samples, with a total of 10 prediction categories, forming a 990×11 matrix. In this research, four syllables were randomly selected from the 11 syllables as the prediction categories, namely 'hid,' 'hEd,' 'hYd,'

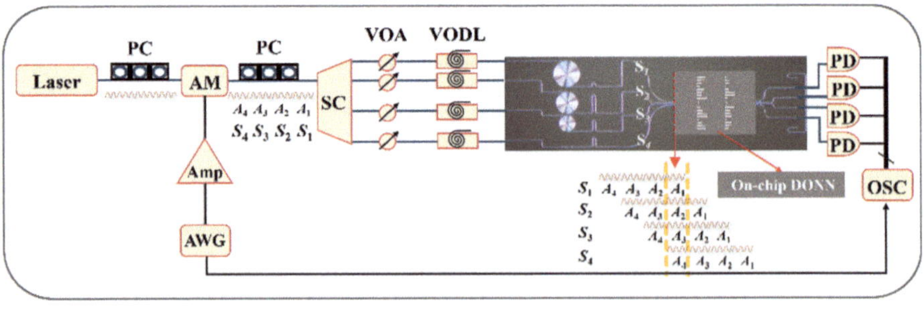

Fig. 4.8 Schematic diagram of the silicon-based integrated DONN chip with serial-to-parallel conversion based on arrayed waveguides [1]

and 'hOd.' The training and testing sets were randomly sampled from the original dataset with 180 samples. Consequently, the resulting dataset is a 360 × 11 matrix, with 180 samples in both the training and testing sets, four prediction categories, and each syllable comprising 10 features. To meet the experimental premise of the 1 × 4 arrayed waveguide, this research work compressed the 10 features of each syllable into 4 features using a fully connected network. Consequently, the new dataset after feature compression is a 360 × 5 matrix, which includes 180 samples each for the training and testing sets, with four prediction categories, and each syllable characterized by four input features. Based on the vowel recognition dataset after feature compression and in combination with the experimentally verified arrayed waveguide, the phase weight parameters pre-trained for the silicon-based integrated DONN are shown in Fig. 4.9a, b, with 70 phase weight parameters on each hidden layer. Figure 4.9c is a schematic diagram of the physical structure of the silicon-based integrated DONN, where the total length of the SSSDG array is 105 μm, and the spacing between the input layer and the first hidden layer, between hidden layers, and between the last hidden layer and the output layer is 300 μm. Figure 4.9d is a schematic diagram of the mathematical logic reasoning process corresponding to Fig. 4.9c, and Eq. 4.2 is the mathematical logic expression of the physical process in Fig. 4.9c.

$$\begin{bmatrix} y_4 \\ y_3 \\ y_2 \\ y_1 \end{bmatrix}^{\mathrm{T}} = \begin{bmatrix} \\ \boldsymbol{w}^{(3)} \\ \\ \end{bmatrix}^{\mathrm{T}}_{4\times 70} \begin{bmatrix} e^{j\phi_{70}^{2}} & 0 & 0 & \cdots & 0 \\ 0 & e^{j\phi_{69}^{2}} & 0 & \cdots & 0 \\ \vdots & \ddots & \ddots & \ddots & \vdots \\ 0 & \cdots & 0 & \ddots & 0 \\ 0 & \cdots & 0 & 0 & e^{j\phi_{1}^{2}} \end{bmatrix} \begin{bmatrix} \\ \boldsymbol{w}^{(2)} \\ \\ \end{bmatrix}^{\mathrm{T}}_{70\times 70} \begin{bmatrix} e^{j\phi_{70}^{1}} & 0 & 0 & \cdots & 0 \\ 0 & e^{j\phi_{69}^{1}} & 0 & \cdots & 0 \\ \vdots & \ddots & \ddots & \ddots & \vdots \\ 0 & \cdots & 0 & \ddots & 0 \\ 0 & \cdots & 0 & 0 & e^{j\phi_{1}^{1}} \end{bmatrix} \begin{bmatrix} \\ \boldsymbol{w}^{(1)} \\ \\ \end{bmatrix}^{\mathrm{T}}_{70\times 4} \begin{bmatrix} x_4 \\ x_3 \\ x_2 \\ x_1 \end{bmatrix} \tag{4.2}$$

In this training stage, the features of the vowel data were mapped onto the amplitude of light and then used as the input signal for the silicon-based integrated DONN to optimize the network structure parameters. Figure 4.10a shows the variation of the loss function during the DONN training process based on the vowel recognition dataset. As the number of iterative optimization steps in the training process increases, the loss function gradually converges to a stable value. After the optimizing the phase weight parameters of the silicon-based integrated DONN was completed, predictions were made for the 180 test set samples in the vowel recognition dataset, with an accuracy rate of 98.3%. Figure 4.10b is the confusion matrix of the test set prediction results. In addition, Fig. 4.10c–d show the prediction results for each type of vowel. The labels in this simulation are referenced from the amplitude, but the intensity of light can also be used as a reference; both methods are viable.

In this work, the silicon-based integrated DONN chip with serial-to-parallel conversion input based on arrayed waveguides can handle high-speed serial signal input tasks. The calculation method for the computing speed of this silicon-based integrated DONN system differs from Eq. 3.1 introduced in Chap. 3. The method presented in Chap. 3 did not account for the time required for the input signal loading

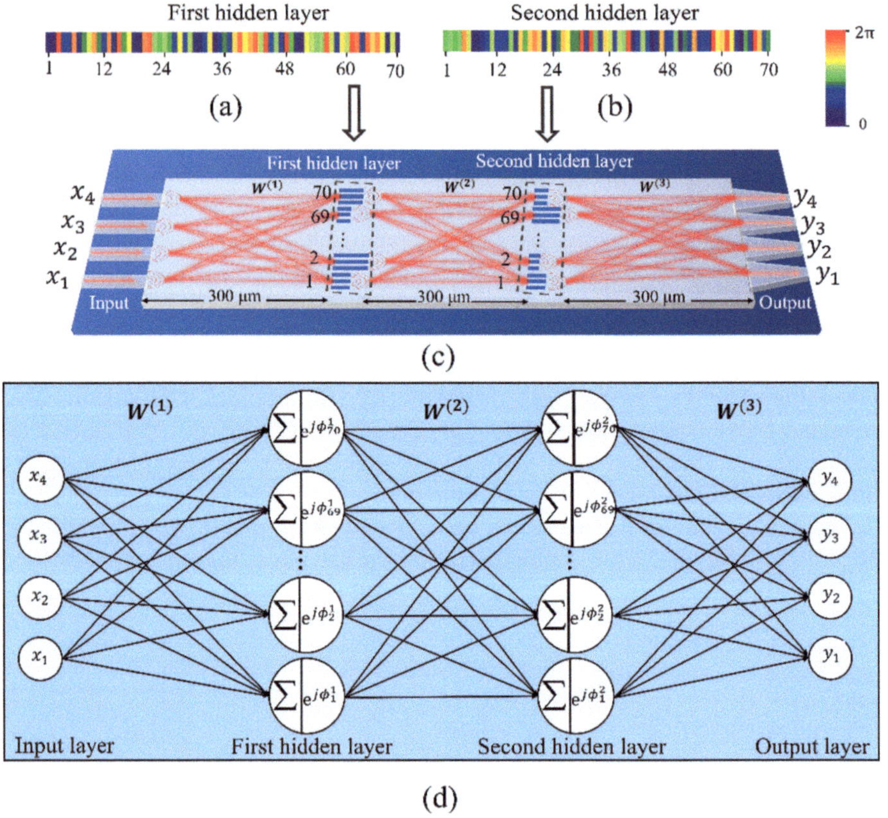

Fig. 4.9 Silicon-based integrated DONN architecture and mathematical logic reasoning demonstration for vowel recognition task [1]. **a** and **b** show the phase weights of the first and second hidden layers of the silicon-integrated DONN, respectively; **c** Schematic diagram of the physical structure of the silicon-based integrated DONN; **d** Mathematical logic diagram of **c**

process. However, in this subsection, the time consumed by the input signal loading process is considered when calculating the computing speed of the silicon-based integrated DONN. The specific calculation method is shown in Eq. 4.3.

$$R = 2m \times N^2 \times \min\left(\frac{f}{2M - 1}, D_{\text{bw}}\right) \text{ OPS} \tag{4.3}$$

where R represents the number of operations completed per second; N denotes the number of computing units on each hidden layer; m indicates the number of $N \times N$ matrices; D_{bw} represents the detection bandwidth of the photodetector; f is the modulation rate of the input signal modulator; M is the number of waveguides in the arrayed waveguide structure, which corresponds to the dimension of the parallel input signals after serial-to-parallel conversion. At this point, the computing efficiency of

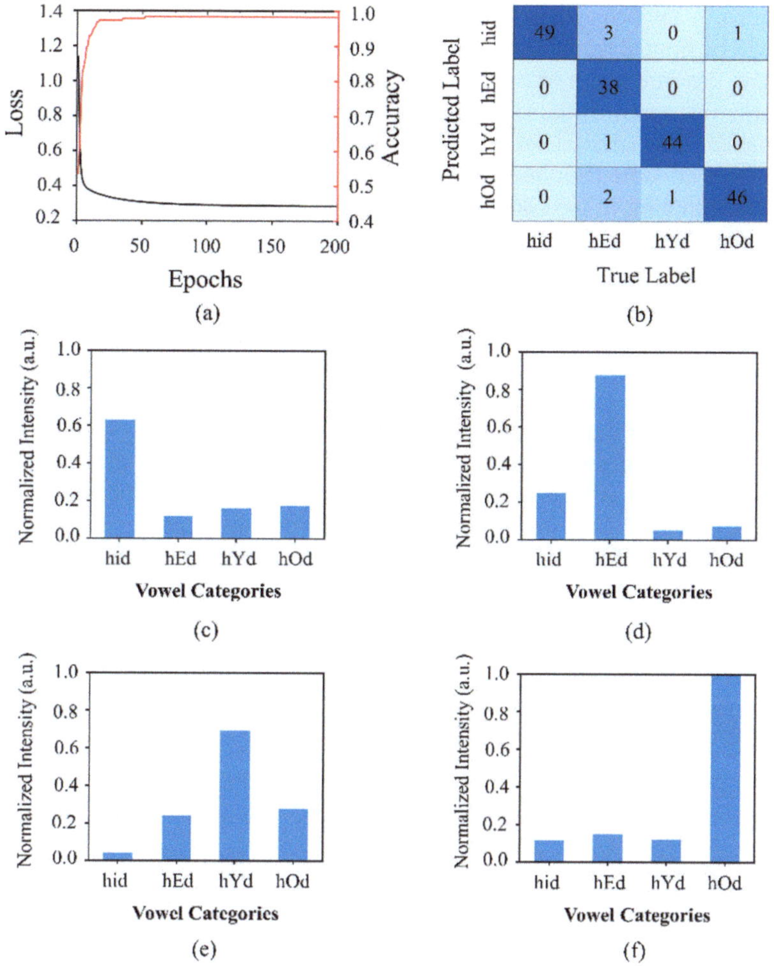

Fig. 4.10 Simulation results of the silicon-based integrated DONN designed for the vowel recognition task [1]. **a** Variation of the loss function during the training process and the blind test accuracy of the test set; **b** Confusion matrix of the test set prediction results; **c–f** Prediction results for the vowel 'hid', 'hEd', 'hYd', and 'hOd', respectively

the silicon-based integrated DONN system is jointly affected by the modulation rate of the input signals, the dimension of the parallel input signals after serial-to-parallel conversion, and the bandwidth of the photodetector.

For example, in this design, the modulation rate of the input signal is 10 Gbps, and the dimension of the parallel input signals after serial-to-parallel conversion is 4. Therefore, in Eq. 4.3, $f/(2M-1) \approx 1.4$ Gbps. Regardless of the value of the detector bandwidth D_{bw}, as long as D_{bw} is greater than 1.4 GHz, the computing speed of the silicon-based integrated DONN system will be determined by the value of $f/(2M-1)$. On the contrary, the computing speed of the DONN system will be determined by the detector bandwidth D_{bw}. In other words, the system's computing speed will be limited by the lower of the two values: the modulation rate of the input signal and the bandwidth of the output signal photodetector. According to Eq. 4.3, the computing speed of the silicon-based integrated DONN system in this design is calculated to be approximately 1.4×10^{13} OPS. The input power of the laser is 32 mW, and all structures on the chip are passive devices. Therefore, the on-chip power consumption of the silicon-based integrated DONN is about 2.286×10^{-17} J/Operation. Additionally, Table 4.2 compares some performance aspects between the silicon-based integrated DONN with serial-to-parallel conversion input based on arrayed waveguides and other silicon-based integrated ONNs.

In Table 4.2, MZI is the abbreviation for Mach–Zehnder interferometer, MMI is the abbreviation for multimode interferometer, SWU is the abbreviation for subwavelength unit, DMD is the abbreviation for digital micromirror devices, and MTPS is the abbreviation for multiple thermo-optic phase shifters. Through comparison, the input signal loading rate of the silicon-based integrated DONN system based on arrayed waveguides is three orders of magnitude higher than that of other silicon-based integrated ONN systems. Moreover, it is capable of processing high-speed serial signal inputs. Its on-chip inference process is entirely passive and all-optical, resulting in low power consumption. Compared with other silicon-based integrated ONNs, the silicon-based integrated DONN chip in this work boasts unique benefits.

Table 4.2 Performance comparison of serial-to-parallel conversion input silicon-based integrated DONN with other silicon-based integrated ONNs

Works	Computing unit	Input signal loading method	Input signal loading rate	On-chip power consumption
Ref. [3]	MZI	MTPS	~ Kbps	1.92 W
Ref. [4]	MZI	MTPS	~ Kbps	3.92 W
Ref. [5]	MZI	MTPS	~ Kbps	7.7 mW
Ref. [6]	MMI	MTPS	~ Kbps	2.04 W
Ref. [7]	SWU	DMD + Lens group	~ Kbps	Passive
Ref. [8]	SWU	MTPS	~ Kbps	120 mW
This work	SWU	High-speed modulator + Arrayed waveguides	~ 1.4 Gbps	Passive

4.3 Multi-task Silicon-Based Integrated Diffractive Optical Neural Network

As discussed in Sect. 4.1, the multifunctional silicon-based integrated DONN architecture mainly consists of two parts. The first part is the serial-to-parallel conversion based on the arrayed waveguides. Section 4.2 experimentally verified the effectiveness of the arrayed waveguides in converting serial input signals into parallel input signals. Based on this experimental premise, a silicon-based integrated DONN chip was designed for the vowel recognition dataset, and the serial-to-parallel input silicon-based integrated DONN performance was validated through numerical calculations. Therefore, Sect. 4.2 has validated the feasibility of inputting serial signals into the silicon-based integrated DONN through high-speed serial-to-parallel conversion by combining experiments and numerical calculations. The second part is the multi-task silicon-based integrated DONN, focusing primarily on its design and implementation. The function of the multi-task silicon-based integrated DONN is to enable a single chip to process multiple different (more than two) tasks. Therefore, compared with the previously studied silicon-based integrated DONNs, the multi-task silicon-based integrated DONN will have a broader range of applications. There are many methods in electronic neural networks to endow a single neural network with the ability to handle multiple tasks, including multi-task learning methods [9–12] and transfer learning methods [13–16]. This section will combine these two methods to implement the multi-task processing capability of the silicon-based integrated DONN and then verify its performance through numerical calculations.

4.3.1 Multi-task Learning

Multi-task learning is a process of learning different tasks simultaneously, where the neural network structure parameters learned through this process can be applied to multiple distinct tasks. This study aims to obtain a set of DONN structure parameters through optimization, which can be shared when processing different tasks. Suppose there are a total of N tasks, with each task t_a corresponding to a dataset $\boldsymbol{D}^{t_a} = \{\boldsymbol{x}_i^{t_a}, y_i^{t_a}\}_{,}$, $(i = 1, 2, \cdots, n_{t_a})$, where n_{t_a} denotes the number of samples in task t_a. For example, if task t_1 is the classification task of the IRIS Plants Dataset, then $\boldsymbol{D}^{t_1}$ represents a 150×5 matrix; $\boldsymbol{x}^{t_1}$ represents a 150×4 matrix, where the 120×4 matrix is the training set and the 30×4 matrix is the test set; while $\boldsymbol{y}^{t_1}$ is a 150×1 vector, composed of 0, 1, and 2, representing the class labels of the iris flowers, respectively.

This research selected the MNIST dataset [17] and the Fashion-MNIST dataset [18] for multi-task learning. The MNIST dataset and its data preprocessing were described in detail in Sect. 3.3. The Fashion-MNIST dataset is a ten-classification task that differs from the MNIST dataset, which primarily classifies ten items: trousers, dresses, coats, shirts, T-shirts, pullovers, sneakers, sandals, ankle boots, and bags.

These categories are labeled with numbers 0 to 9. The training set of Fashion-MNIST contains 60,000 images, while the test set contains 10,000 images. Each image is 28 × 28 pixels, the same feature size as the MNIST dataset. Therefore, the data must be preprocessed before designing the silicon-based integrated DONN chip based on this dataset. First, the 28 × 28-pixel images are converted into a 784 × 1 vector. Then, through a fully connected network, the 784 × 1 vector is further compressed into a 10 × 1 vector. As a result, the input feature size for the silicon-based integrated DONN chip is 10. Currently, the preprocessed Fashion-MNIST dataset is similar to the preprocessed MNIST dataset. Both have a training set with a 60,000 × 10 matrix and a test set with a 10,000 × 10 matrix. The training and test set labels are 60,000 × 1 and 10,000 × 1 vectors, respectively.

Before optimizing the structure parameters of the multi-task silicon-based integrated DONN, this work concatenated the MNIST dataset and the Fashion-MNIST dataset into a larger dataset, namely the multi-task dataset $T_a = \{t_{\text{MNIST}}, t_{\text{Fashion-MNIST}}\}$. The new dataset is denoted as $\boldsymbol{D}^{T_a} = \left\{\boldsymbol{x}_i^{T_a}, y_i^{T_a}\right\}$ $(i = 1, 2, ..., n_{T_a})$, where $\boldsymbol{D}^{T_a}$ represents a 140,000 × 11 matrix; $\boldsymbol{x}^{T_a}$ represents a 140,000 × 10 matrix, with a 120,000 × 10 matrix serving as the training set and a 20,000 × 10 matrix as the test set; while $\boldsymbol{y}^{T_a}$ is a 140,00 × 1 vector, composed of numbers from 0 to 9, representing the class labels for the new combined task.

The multi-task T_a includes two tasks, namely task t_{MNIST} and task $t_{\text{Fashion-MNIST}}$. The training process can be referred to as shown in Fig. 4.11, where $f(\cdot)$ is the mapping function that needs to be learned for the multi-task T_a. The objective of learning this mapping function is to minimize the value of the loss function $L\left(f\left(\boldsymbol{x}_i^{T_a}\right), y_i^{T_a}\right)$, and the expression for L is given in Eq. 4.4.

$$L = \frac{1}{n}\sum_{i=1}^{n}\left(f\left(\boldsymbol{x}_i^{T_a}\right) - y_i^{T_a}\right)^2 \tag{4.4}$$

where L is the loss function to be optimized, n is the total number of samples of task T_a, $\boldsymbol{x}_i^{T_a}$ represents the i-th sample in task T_a, $f\left(\boldsymbol{x}_i^{T_a}\right)$ represents the output result of the i-th sample in task T_a after being calculated by the mapping function $f(\cdot)$, and $y_i^{T_a}$ represents the label of the i-th sample in task T_a.

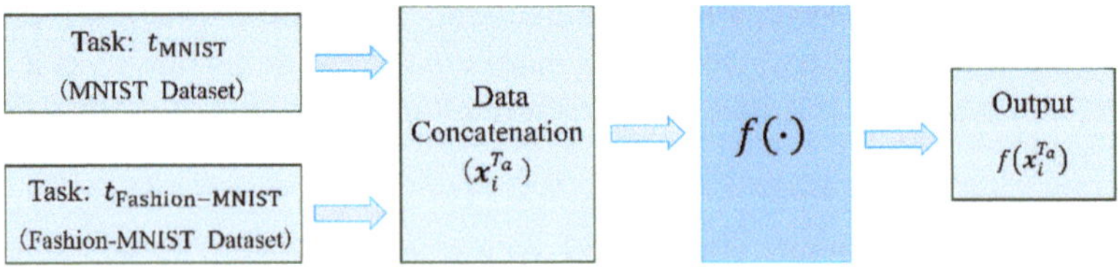

Fig. 4.11 Schematic diagram of the shared-weight training process for the MNIST dataset and the Fashion-MNIST dataset

4.3.2 Transfer Learning

Transfer learning is at least concerned with two or more tasks, where the dataset being transferred belongs to the source domain data, and the network being transferred is called the source domain network. The dataset to be transferred belongs to the target domain data, and the network that completes the target task is called the target domain network. The MNIST dataset and the USPS dataset [19] are datasets for transfer learning. Among them, the USPS dataset is also a dataset for a ten-classification task, mainly for classifying and recognizing the digits 0–9 on stamps. The USPS dataset used in this study consists of 11,000 images, with 7700 images in the training set and 3,300 images in the test set. Each image is 16 × 16 pixels in size. Before using this dataset, data preprocessing is also required. First, the 16 × 16-pixel images are converted into 256 × 1 vectors. Since 256 features are still too many and the silicon-based integrated DONN chip struggles to handle such high-dimensional input signals, this study further compresses the 256 × 1 vector into a 10 × 1 vector using a fully connected network. As a result, the input feature size for the silicon-based integrated DONN chip is reduced to 10. The USPS training set forms a 7700 × 10 matrix, and the test set forms a 3300 × 10 matrix. The labels for the USPS dataset are a 11,000 × 1 vector, composed of numbers ranging from 0 to 9.

As shown in Fig. 4.12, the source domain network is the network structure trained on the MNIST dataset. In contrast, the target domain network is obtained by retraining the output layer parameters of the optimized source domain network. This network can handle the classification task of the USPS dataset. From Fig. 4.12, the source domain network optimized on the MNIST dataset has all its parameters, except for the output layer, copied to the target domain network. The only difference is that the output layer parameters in the target domain network are randomly initialized and retrained. In fact, the target domain network requires the retraining of very few parameters, which saves a significant amount of training resources compared to the source domain network. The application value of transfer learning is more critical in some special cases, such as when applied between multiple large datasets, where it can save a substantial amount of computing resources.

4.3.3 Design of Structures

This section first creates a new dataset by concatenating the MNIST and Fashion-MNIST datasets. Then, the structure parameters of the silicon-based integrated DONN system are trained based on this new dataset, as shown in Fig. 4.13. There are 10 input waveguides, with a center-to-center spacing of 6 μm between adjacent input waveguides. The cascaded waveguides between DONN1 and DONN2 and the output waveguides consist of 10 waveguides, with a spacing of 8 μm between adjacent waveguides. The basic computing unit in the silicon-based integrated DONN system, namely the subwavelength diffractive unit, is still implemented using SSSDG. Each

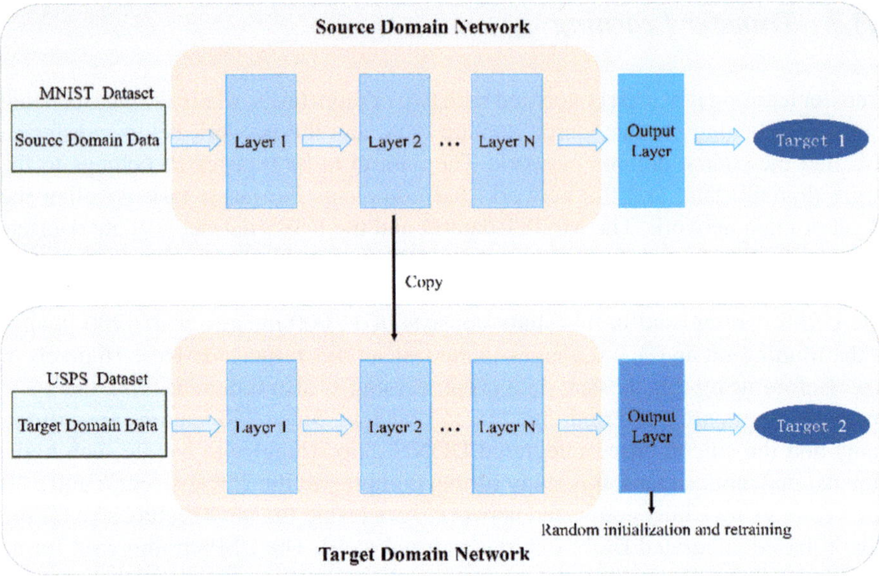

Fig. 4.12 Schematic diagram of the transfer learning process for the MNIST dataset and the USPS dataset

SSSDG is composed of three identical SSSDs. Each SSSD has a width of 200 nm, a height of 220 nm, and a longitudinal (y-axis) arrangement period of 500 nm. The length of each SSSD is distributed between 0.2 μm and 2.5 μm. Therefore, the longitudinal arrangement period (width) of the SSSDG is 1.5 μm, and its length is also distributed between 0.2 μm and 2.5 μm. Each hidden layer (SSSDG array) in the silicon-based integrated DONN contains 60 SSSDGs (computing units), making the length of the SSSDG array 90 μm.

In Fig. 4.13, phase shifters are arranged on both the input layer waveguides and the cascaded waveguides between DONN1 and DONN2. The phase shifters on the input layer waveguides are used for input signal loading and system error compensation,

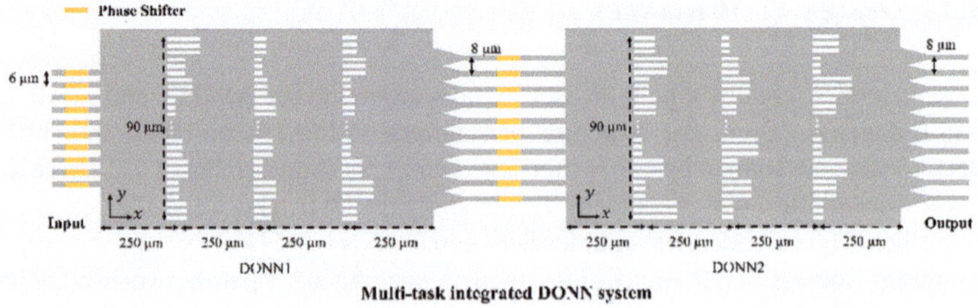

Fig. 4.13 Schematic diagram of the multitask silicon-based integrated DONN system structure

while the phase shifters on the cascaded waveguides between DONN1 and DONN2 are specifically set for system error compensation. The use of these phase shifters on the cascaded waveguides between DONN1 and DONN2, including whether to use them and how many to use, can be determined based on the specific circumstances of the experimental system errors. The phase weight parameters on each hidden layer are pre-trained on a computer. The input dataset for this training is the new dataset concatenated from the MNIST dataset and the Fashion-MNIST dataset. This dataset consists of 120,000 training samples and 20,000 test samples, with 140,000 labels. For more details on the new dataset, refer to Sect. 4.3.1.

In this research, based on the structure shown in Fig. 4.13, a silicon-based integrated DONN hyperparameter setup was constructed on a computer, and the optimization of the phase weight parameters on each hidden layer of the multi-task silicon-based integrated DONN system began under this hyperparameter configuration. As shown in Fig. 4.14, the variation of the loss function with the number of optimization training iterations indicates that as the number of optimizations increases, the value of the loss function gradually converges and stabilizes around 0.015. At this point, the structure parameters of the multi-task silicon-based integrated DONN system have been obtained through optimization.

This research aims to obtain a multi-task silicon-based integrated DONN system structure that, after being optimized in training, can handle both the MNIST and Fashion-MNIST classification tasks simultaneously. Once the structure parameters of this multi-task silicon-based integrated DONN system are obtained and fixed, given that the MNIST and USPS datasets meet the conditions for transfer learning, an additional output layer is appended to the DONN system. The parameters of this new output layer are then randomly initialized and retrained. The newly trained output layer, combined with the previously trained silicon-based integrated DONN system, can then be used to predict the categories of the USPS dataset. It is worth noting that the input data for training the new output layer (fully connected layer) parameters are the output optical power values from the multi-task silicon-based integrated

Fig. 4.14 Variation of the loss function during the optimization of parameters for the multi-task silicon-based integrated DONN

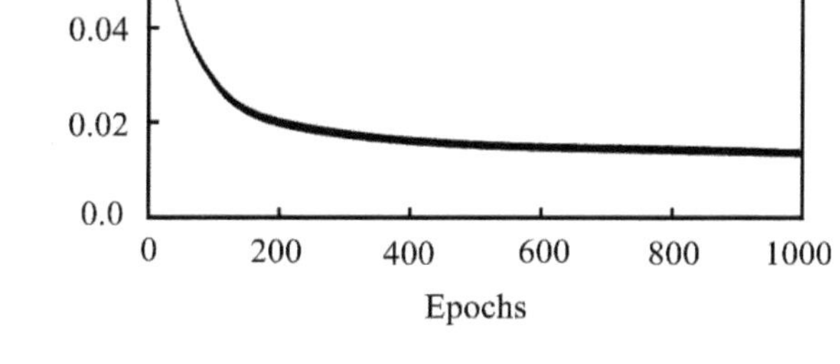

DONN system. After obtaining these optical power values, they are used as a new dataset to complete the training of the output fully connected layer parameters via a computer. Additionally, during the training process of the output fully connected layer parameters, only the training set from the USPS dataset is involved, while the test set is not used for training. Once the new output fully connected layer parameters are trained on the computer, the USPS dataset can be classified based on the multi-task silicon-based integrated DONN system transfer network (target domain network) shown in Fig. 4.15.

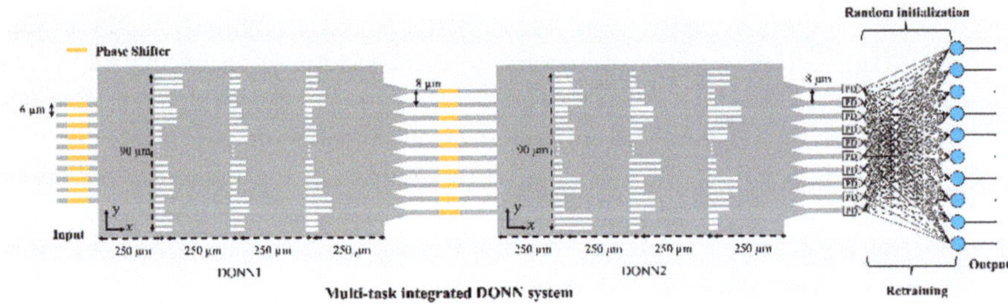

Fig. 4.15 Schematic diagram of the transfer structure of the multi-task silicon-based integrated DONN system

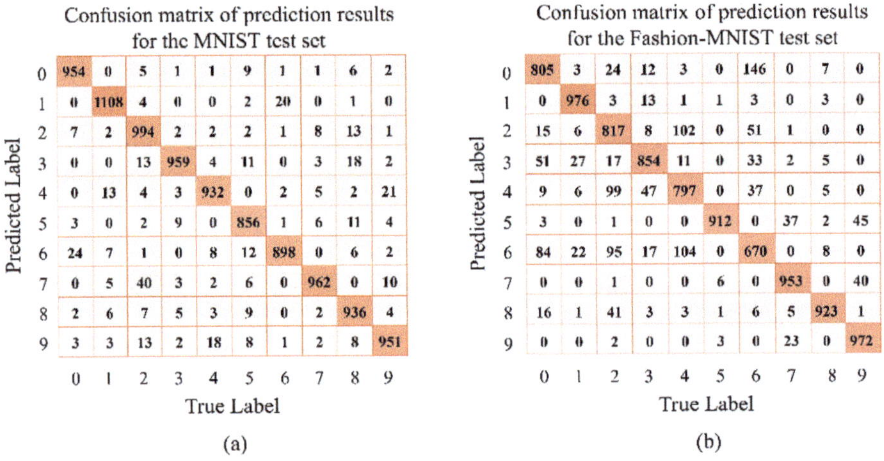

Fig. 4.16 Prediction results of the multi-task silicon-based integrated DONN system for the MNIST and Fashion-MNIST test sets. **a/b** Confusion matrix of the test results for the MNIST/ Fashion-MNIST test set

4.3.4 Simulation Verification

In Sect. 4.3.3, the structure parameters of the multi-task silicon-based integrated DONN system were obtained through optimization. This subsection will verify the performance of the multi-task silicon-based integrated DONN system using the test sets from the MNIST and Fashion-MNIST datasets. Figures 4.16a and b are the confusion matrices of the test results for the MNIST test set and the Fashion-MNIST test set, respectively.

In Sect. 4.3.3, the training of the transfer structure for the new USPS dataset was completed based on the existing structure parameters of the silicon-based integrated DONN system. This involved adding a new output layer after the original output structure, a fully connected layer whose parameters must be retrained using machine learning algorithms. Figure 4.17 shows the variation of the loss function during the training process of the output fully connected layer parameters. Additionally, Fig. 4.17 illustrates the change in the blind test accuracy of the USPS test set as the output fully connected layer parameters are continuously optimized and updated. It can be seen from the figure that as the number of optimization iterations increases, the loss function keeps decreasing and tends to stabilize. At the same time, the blind test accuracy of the USPS test set continues to rise. Furthermore, Fig. 4.18 presents the confusion matrix of the prediction results for the USPS test set by the multi-task silicon-based integrated DONN system after training the output fully connected layer parameters.

The prediction results of the multi-task silicon-based integrated DONN system for the MNIST dataset, Fashion-MNIST dataset, and USPS dataset are shown in Table 4.3. These results are comparable to, and in some cases slightly higher than, those achieved by the current state-of-the-art specialized ONNs [4, 5, 7, 8, 20] for the same or similar tasks. This indicates that the multi-task silicon-based integrated DONN system can perform well when dealing with multiple tasks.

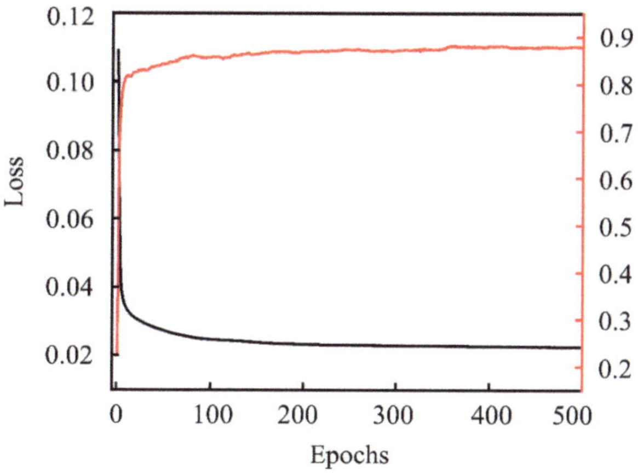

Fig. 4.17 Variation of the loss function during the training of the output fully connected layer (new output layer) parameters of the multi-task silicon-based integrated DONN system, and the display of the blind test accuracy results for the USPS test set

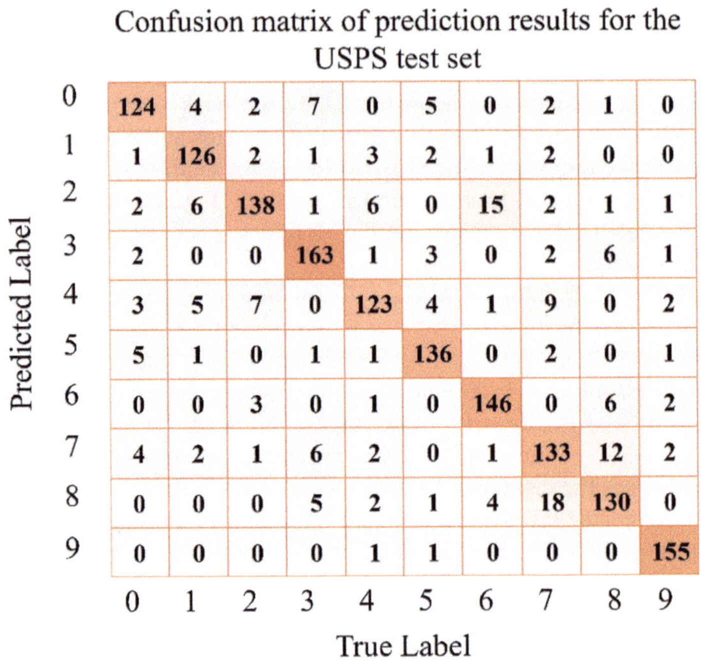

Fig. 4.18 Prediction results of the multi-task silicon-based integrated DONN system for the USPS test set

Table 4.3 Classification results for different tasks by the multi-task silicon-based integrated DONN system

Task (datasets)	Classification accuracy on the test set (%)
MNIST	95.1
Fashion-MNIST	87.3
USPS	87.7

4.4 Summary

This chapter addresses the limitation of the silicon-based integrated DONN chip designed solely for a single task in Chap. 3 by proposing a multifunctional silicon-based integrated DONN system architecture. This architecture mainly consists of two parts. The first part is designed based on the arrayed waveguides to enable the input section of the silicon-based integrated DONN chip to achieve high-speed serial-to-parallel conversion. The second part incorporates multi-task learning and transfer learning methods to endow a single silicon-based integrated DONN chip with the ability to handle multiple tasks. Integrating these two parts can realize a multifunctional silicon-based integrated DONN system. This system is capable of processing various tasks with high-speed serial signal inputs. Moreover, as the modulation rate of the input signal increases, the computing efficiency and processing capability of this multifunctional silicon-based integrated DONN will be further enhanced.

In Sect. 4.1 of this chapter, the network architecture of the multifunctional silicon-based integrated DONN was explicitly introduced. Section 4.2 experimentally verified the excellent performance of the serial-to-parallel conversion technology. Based on the serial-to-parallel conversion experiment, the classification effect of the silicon-based integrated DONN designed with this technology on the vowel recognition dataset was validated through numerical calculations. Additionally, in Sect. 4.3, a multi-task silicon-based integrated DONN system was designed. This system can handle the classification tasks of multiple datasets, including MNIST, Fashion-MNIST, and USPS, under fixed phase weight parameters. The classification accuracy is comparable to that of other state-of-the-art ONNs and is even slightly higher than the classification accuracy of some ONNs. The experimental verification and numerical calculation results in Sects. 4.2 and 4.3 demonstrate the feasibility of realizing a multifunctional silicon-based integrated DONN chip. Once the optimization of the entire multifunctional silicon-based integrated DONN system structure parameters is completed, the length of the corresponding SSSDG for each phase weight on the hidden layers can be calculated using Eq. 2.14. Subsequently, the phase weights pre-trained on the computer can be accurately mapped onto the physical structure of the chip through the phase weight mapping model. As for the subsequent chip fabrication process, the method is the same as that introduced in Chap. 3. Therefore, the preparation of the multifunctional silicon-based integrated DONN chips can be completed in the same manner.

The network architecture and design methodology of the multifunctional silicon-based integrated DONN chip proposed in this chapter enable the chip to handle high-speed serial signal inputs and endow it with the capability to process multiple distinct tasks. Compared to the silicon-based integrated DONN chip discussed in Chap. 3, these advantages represent significant progress in both computing efficiency and various applications, and they will become more pronounced as the modulation rate of the input signal increases.

References

1. Fu T, Huang Y, Sun R et al (2023) Integrated diffractive optical neural network with space-time interleaving. Chin Opt Lett 21(9):091301
2. Blake C. UCI repository of machine learning databases. http://www.ics.uci.edu/~mlearn/MLRepository.html, 1998
3. Shen Y, Harris NC, Skirlo S et al (2017) Deep learning with coherent nanophotonic circuits. Nat Photonics 11(7):441–446
4. Zhang H, Gu M, Jiang XD et al (2021) An optical neural chip for implementing complex-valued neural network. Nat Commun 12(1):457
5. Zhu HH, Zou J, Zhang H et al (2022) Space-efficient optical computing with an integrated chip diffractive neural network. Nat Commun 13(1):1044
6. Zhao X, Lv H, Chen C et al (2021) On-chip reconfigurable optical neural networks

7. Wang Z, Chang L, Wang F et al (2022) Integrated photonic metasystem for image classifications at telecommunication wavelength. Nat Commun 13(1):2131
8. Fu T, Zang Y, Huang Y et al (2023) Photonic machine learning with on-chip diffractive optics. Nat Commun 14(1):70
9. Xue Y, Liao X, Carin L et al (2007) Multi-task learning for classification with dirichlet process priors. J Mach Learn Res 8(1)
10. Chen JYC, Terrence PI (2009) Effects of imperfect automation and individual differences on concurrent performance of military and robotics tasks in a simulated multitasking environment. Ergonomics 52(8):907–920
11. Li H, Liao X (2009) Multi-task reinforcement learning in partially observable stochastic environments. J Mach Learn Res 10(5)
12. Thung KH, Wee CY (2018) A brief review on multi-task learning. Multimedia Tools Appl 77(22):29705–29725
13. Torrey L, Shavlik J (2010) Transfer learning. Handbook of research on machine learning applications and trends: algorithms, methods, and techniques. IGI Global 242–264
14. Lu Z, Zhu Y, Pan S et al (2014) Source free transfer learning for text classification. In: Proceedings of the AAAI conference on artificial intelligence 28(1)
15. Zhuang F, Qi Z, Duan K et al (2020) A comprehensive survey on transfer learning. Proc IEEE 109(1):43–76
16. Panigrahi S, Nanda A, Swarnkar T (2021) A survey on transfer learning. intelligent and cloud computing: proceedings of ICICC 2019, vol 1. Springer Singapore, pp 781–789
17. LeCun Y (1998) The MNIST database of handwritten digits. http://yann.lecun.com/exdb/mnist/
18. Xiao H, Rasul K, Vollgraf R Fashion-mnist: a novel image dataset for benchmarking machine learning algorithms. ArXiv preprint arXiv:1708.07747
19. Hull JJ (1994) A database for handwritten text recognition research. IEEE Trans Pattern Anal Mach Intell 16(5):550–554
20. Lin X, Rivenson Y, Yardimci NT et al (2018) All-optical machine learning using diffractive deep neural networks. Science 361(6406):1004–1008

Chapter 5
Conclusions and Future Perspectives

Abstract This chapter systematically summarizes the research on the silicon-based integrated DONNs, provides an outlook on the unresolved key scientific challenges, and offers constructive suggestions for addressing the technical issues that remain to be solved.

5.1 Conclusions of Research Work

This thesis starts with research on the silicon-based integrated DONN model and then conducts a theoretical exploration of two-dimensional silicon-based SDEPM. Inspired by the Huygens-Fresnel diffraction formula, an analytical model for the optical field evolution of a silicon-based slab waveguide with specific size was established, and based on this, the two-dimensional silicon-based SDEPM was obtained. In the research on the silicon-based integrated DONN model, the effective refractive index of the subwavelength micro-slot region with different sizes is calculated using the EMT and 2.5D FDTD methods. Based on this, a phase weight mapping model is proposed under an approximate condition. Then, based on the research of the silicon-based integrated DONN model, a silicon-based integrated DONN chip was designed for a specific classification task. Additionally, an effective error compensation scheme is proposed for the system errors arising during the fabrication and experimental testing of the silicon-based integrated DONN chip. Since the silicon-based integrated DONN chip is designed for a specific task, it cannot handle multiple tasks. Furthermore, the silicon-based integrated DONN chips developed in this study utilize thermo-optic phase shifters for parallel input signal loading, which inherently limits the signal loading rate due to the relatively slow response speed of the thermo-optic effect. Consequently, this research has also conducted relevant studies on the multifunctional silicon-based integrated DONN system.

The results of this thesis are compared with the research objectives proposed in the introduction part, as shown in Table 5.1. The computing energy (0.152 W) includes the input power of the laser (32 mW) and the energy consumption required for the operation of four thermo-optic phase shifters (120 mW). This power represents

© Tsinghua University Press 2026

T. Fu, *Research on Silicon-Based Integrated Diffractive Optical Neural Network*, Springer Theses, https://doi.org/10.1007/978-981-95-3787-7_5

the energy the photonic chip consumes to complete one parallel computation. If calculated based on the energy consumed per operation, the power consumption is approximately 1.1×10^{-17} J/Operation.

In Table 5.1, the performance parameters of the silicon-based integrated DONN-I3 chip are compared with the research goals proposed in the introduction, where each computing unit consists of three identical SSSDs. The minimum area of the SSSD is 0.1 μm^2, and the maximum area is 1.25 μm^2, so the computing unit SSSDG in this study is approximately 3 μm^2. Its computing energy efficiency is about 1.1×10^{-17} J/Operation, which is 2–3 orders of magnitude lower than the energy consumption of other silicon-based integrated ONNs at present, thus meeting the requirements of the research goals. As for its computing speed (1.38×10^{16} OPS), it is already four orders of magnitude higher than the research goal of $\sim 10^{12}$ OPS (i.e. the advanced computing speed of electronic computing hardware at present [1]). Furthermore, the silicon-based integrated DONN-I3 chip has the advantage of on-chip integration, which has good stability and portability, and its computing power consumption (1.1×10^{-17} J/Operation) is also very low. Therefore, all the parameters and indicators mentioned in the introduction of the research goals have been achieved in this thesis.

The main innovations of the research work in this thesis are summarized as follows:

(1) A two-dimensional silicon-based SDEPM and a phase weight mapping model were proposed. Among them, the two-dimensional silicon-based SDEPM was inspired by the Huygens-Fresnel principle in the transmission medium of silicon-based slab waveguides. This model can approximate the propagation laws of light in silicon-based slab waveguides analytically, and based on this, the forward propagation model in the training of silicon-based integrated DONN can be analytically established on a computer. Thus, the structure parameters of silicon-based integrated DONN can be trained efficiently. The phase weight mapping model not only solves the problem of uneven medium distribution in the subwavelength micro-slot region but also solves the problem of unstable effective refractive index calculation of incident light at different incidence angles, making the mapping process of pre-trained silicon-based integrated DONN structure parameters to physical devices more accurate.

Table 5.1 Research objectives of this thesis

Parameters	Dimension	Size of the computing unit	Theoretical integration degree (computing units/mm^2)	Energy efficiency	Stability/ portability	Computing speed (operations/ s)
Objective	Integrated	$\leq 1.5\mu m \times 2\ \mu m$	$\backsim 10^3$	High	Good	$>10^{12}$
Research findings (DONN-I3)	Integrated	$\backsim 3\mu m^2$	$\backsim 2 \times 10^3$	0.152 W	Good	$\backsim 1.38 \times 10^{16}$

(2) A silicon-based integrated DONN chip with low latency, low power consumption, high integration, and high computing capacity was proposed. The DONN chip's structure parameters (weights) were pre-trained on a computer, and subwavelength diffractive units realized the weights. The computing process was completed through the diffraction and interference of light, which ensures that the computing power consumption will not increase with the large-scale expansion of the computing units. Due to unavoidable process errors in chip fabrication and additional system errors brought by chip packaging and experimental testing, this study proposed a practical and feasible system error compensation scheme for the silicon-based integrated DONN chip system, significantly improving the system's robustness. For example, the blind test accuracy of the IRIS test set for the DONN-I1 and DONN-I3 chips reached 86.7% and 90.0%, respectively, consistent with the theoretical simulation results after system error compensation.

(3) A design method for a multifunctional silicon-based integrated DONN system is proposed, which can meet the requirements of high-speed serial signal input and simultaneously process multiple tasks. The high-speed serial signal input is realized through a silicon-based arrayed waveguide structure, and as the modulation rate of the input signal increases, the spacing between adjacent waveguide delay lines becomes shorter, and the time required for complete signal serial-to-parallel conversion is significantly reduced. The ability to process multiple tasks is achieved by using multi-task learning and transfer learning methods during the training of the structure parameters of the silicon-based integrated DONN system. In summary, the silicon-based integrated DONN system designed by the proposed method can convert high-speed serial input signals to parallel signals. It can process multiple tasks simultaneously, further expanding the applications of DONN chips and improving their computing efficiency.

5.2 Future Perspectives of Research Work

This research has significantly progressed in developing silicon-based integrated DONN chips; however, several promising directions warrant further investigation. The main research avenues for future exploration include:

1. The phase weight mapping model used in this research was constructed under the approximate condition, so the entire system's design is constrained. In the future, methods such as inverse design and equivalent fitting can be used to obtain a more optimal phase weight mapping model.
2. This research has achieved the capability of the silicon-based integrated DONN chip to handle multiple tasks, but it has not addressed the reconfigurability issue of the DONN computing units. Therefore, the reconfigurability issue of future silicon-based integrated DONN chips could be addressed by replacing the filling material in the subwavelength rectangular trenches (e.g., replacing silicon dioxide material with PCM [2]). Still, the new problem introduced is that the fabrication

process is more challenging. In addition, it can be considered to combine other materials with strong reconfigurability and integrated fabrication platforms to achieve the reconfigurable function of the integrated photonic chip, such as recent reports on reconfigurable photonic integrated chips based on lithium niobate [3] processing platform and indium gallium arsenide phosphide [4] processing platform.

3. The implementation of nonlinearity is a significant challenge for ONNs. This research work implemented nonlinearity only at the output layer using optical detectors. However, no nonlinear functionality was introduced after each hidden layer in the silicon-based integrated DONN chip. As a result, at this stage, the silicon-based integrated DONN chip has only managed to solve some simple linear tasks. However, the difficulty of directly applying optical nonlinearity to ONNs is significant, and the key to breakthroughs lies in developing special materials. Additionally, ONNs that are electrically assisted can implement nonlinearity in the electrical domain. Therefore, a hybrid approach combining photonics and electronics may be a viable solution to address the nonlinearity issue in ONNs in the future. Specifically, this approach leverages the natural advantages of light to handle large-scale linear matrix operations while delegating the less computationally intensive nonlinear functions to be processed by electrical assistance. However, the practical application of hybrid photonic-electronic computing at this stage is still limited by application scenarios and the efficiency of optoelectronic conversion interfaces (computing speed and energy consumption). The future practical application of hybrid photonic-electronic computing may need to rely on specific scenarios to demonstrate its advantages.
4. The limitation of high-dimensional signal input is a common issue for all two-dimensional silicon-based integrated ONN chips. To address this problem, optical devices with dimensionality reduction capabilities (both two-dimensional and three-dimensional) should be designed to preprocess high-dimensional input signals. Then, the issue of limited high-dimensional signal input can be effectively resolved by reducing the dimensionality of the input data optically and then feeding the reduced data into the two-dimensional silicon-based integrated ONN chips. Since optical reduction is a passive process, implementing dimensionality reduction optically can significantly save the energy consumption required for dimensionality reduction through electrical means.
5. Silicon-based integrated ONN chips possess high computing efficiency. However, due to the limitations of energy efficiency in the optoelectronic (or electro-optic) conversion process, such as at the optoelectronic interface, silicon-based integrated ONN chips currently cannot fully realize their potential in practical applications. Therefore, research into the energy efficiency (computing speed and energy consumption) of the optoelectronic (or electro-optic) conversion process may be extremely necessary.

References

1. Williamson IAD, Hughes TW, Minkov M et al (2019) Reprogrammable electro-optic nonlinear activation functions for optical neural networks. IEEE J Sel Top Quantum Electron 26(1):1–12
2. Delaney M, Zeimpekis I, Du H et al (2021) Nonvolatile programmable silicon photonics using an ultralow-loss Sb2Se3 phase change material. Sci Adv 7(25):eabg3500
3. Yue G, Li Y (2023) Integrated lithium niobate optical phased array for two-dimensional beam steering. Opt Lett 48(14):3633–3636
4. Wu T, Menarini M, Gao Z et al (2023) Lithography-free reconfigurable integrated photonic processor. Nat Photonics 17(8):710–716

MIX
Papier aus verantwortungsvollen Quellen
Paper from responsible sources
FSC® C105338

If you have any concerns about our products,
you can contact us on
ProductSafety@springernature.com

In case Publisher is established outside the EU,
the EU authorized representative is:
Springer Nature Customer Service Center GmbH
Europaplatz 3, 69115 Heidelberg, Germany

Printed by Libri Plureos GmbH
in Hamburg, Germany